W9-DAI-878

Michaeli / Greif / Wolters / Vossebürger

Training in Plastics Technology

Hanser Books Sponsored by SPE

Belofsky, Plastics: Product Design and Process Engineering
Bonenberger, The First Snap-Fit Handbook
Brostow/Corneliussen, Failure of Plastics
Chan, Polymer Surface Modification and Characterization
Charrier, Polymeric Materials and Processing-Plastics, Elastomers and Composites
Chung, Extrusion of Polymers
Del Vecchio, Understanding Design of Experiments: A Primer for Technologists
Ehrig, Plastics Recycling
Ezrin, Plastics Failure Guide
Gordon, Total Quality Process Control for Injection Molding
Gruenwald, Plastics: How Structure Determines Properties
Jones, Guide to Short Fiber Reinforced Plastics
Lee, Blow Molding Design Guide
Macosko, Fundamentals of Reaction Injection Molding
Malloy, Plastic Part Design for Injection Molding
Matsuoka, Relaxation Phenomena in Polymers
Menges/Mohren, How to Make Injection Molds
Michaeli, Extrusion Dies for Plastics and Rubber
Michaeli/Greif/Wolters/Vossebürger, Training in Plastics Technology, 2nd Ed.
Michaeli/Greif/Kretzschmar/Kaufmann/Bertuleit, Training in Injection Molding
Neuman, Experimental Strategies for Polymer Scientists and Plastics Engineers
Osswald, Polymer Processing Fundamentals
Progelhof/Throne, Polymer Engineering Principles
Rauwendaal, Polymer Extrusion
Rees, Mold Engineering
Rosato, Designing with Reinforced Composites
Rotheiser, Joining of Plastics
Saechtling, International Plastics Handbook for the Technologist, Engineer and User
Stevenson, Innovation in Polymer Processing: Molding
Stoeckhert, Mold-Making Handbook for the Plastics Engineer
Tucker, Fundamentals of Computer Modeling for Polymer Processing
Ulrich, Introduction to Industrial Polymers
Wright, Injection/Transfer Molding of Thermosetting Plastics
Wright, Molded Thermosets: A Handbook for Plastics Engineers, Molders and Designers

Michaeli / Greif / Wolters / Vossebürger

Training in Plastics Technology

A Text- and Workbook

2nd Edition

HANSER

Hanser Publishers, Munich

Hanser Gardner Publications, Inc., Cincinnati

The Authors:
Professor Dr.-Ing. Walter Michaeli and *Dipl.-Ing. Leo Wolters*, IKV, Pontstraße 49, 52062 Aachen, Germany; *Dr. phil. Helmut Greif*, Hans-Heyden-Straße 199, 52134 Herzogenrath, Germany; *Dipl.-Ing. Franz-Josef Vossebürger*, Spenglerstraße 3, 90433 Nürnberg, Germany

Original title:
Technologie der Kunststoffe, 2. Auflage, by Michaeli / Greif / Wolters / Vossebürger
© 1998 Carl Hanser Verlag München Wien

Distributed in the USA and in Canada by
Hanser Gardner Publications, Inc.
6600 Clough Pike, Cincinnati, Ohio 45244-4090, USA
Fax: (513) 527-8950
Phone: (513) 527-8977 or 1-800-950-8977
Internet: http://www.hansergardner.com

Distributed in all other countries by
Carl Hanser Verlag
Postfach 86 04 20, 81631 München, Germany
Fax: +49 (89) 98 12 64
Internet: http://www.hanser.de

The use of general descriptive names, trademarks, etc., in this publication, even if the former are not especially identified, is not to be taken as a sign that such names, as understood by the Trade Marks and Merchandise Marks Act, may accordingly be used freely by anyone.

While the advice and information in this book are believed to be true and accurate at the date of going to press, neither the authors nor the editors nor the publisher can accept any legal responsibility for any errors or omissions that may be made. The publisher makes no warranty, express or implied, with respect to the material contained herein.

Library of Congress Cataloging-in-Publication Data
Technologie der Kunststoffe. English
Training in plastics technology : a text- and workbook / [Walter] Michaeli ··· [et al.]
2nd ed.
p. cm.
ISBN 1-56990-293-3 (softcover)
1. Plastics. 2. Polymers. I. Michaeli, Walter. II. Title.
TA455.P5T44313 2000
668.4–dc21 99-048717

Die Deutsche Bibliothek – CIP-Einheitsaufnahme
Training in plastics technology : a text- and workbook / Michaeli
[Transl. by Randall T. Wert]. 2. ed. – Munich ; Vienna ; New York
Hanser ; Cincinnati : Hanser / Gardner, 2000
(SPE books)
Dt. Ausgabe u.d.T.: Technologie der Kunststoffe
ISBN 3-446-21344-9

All rights reserved. No part of this book may be reproduced or transmitted in any form or by any means, electronic or mechanical, including photocopying or by any information storage and retrieval system, without permission in writing from the publisher.

© Carl Hanser Verlag, Munich 2000
Production Management: Martha Kürzl, Stafford UK
Typeset in Great Britain by Techset Composition Ltd., Salisbury
Printed and bound in Germany by Kösel, Kempten

Foreword

The Society of Plastics Engineers is pleased to sponsor and endorse "Training in Plastics Technology." This presentation is intended to provide an uncomplicated and clearly written, yet comprehensive text book covering many major topics associated with plastics technology. The workbook and text with test questions and answers should make an excellent aid to students and technicians.

SPE, through its Technical Volumes Committee, has long sponsored books on various aspects of plastics. Its involvement has ranged from identification of needed volumes and recruitment of authors to peer review and approval and publication of new books.

Technical competence pervades all SPE activities, not only in the publication of books, but also in other areas such as sponsorship of technical conferences and educational programs. In addition, the Society publishes periodicals including – *Plastics Engineering, Polymer Engineering and Science, Polymer Processing and Rheology, Journal of Vinyl Technology and Polymer Composites* – as well as conference proceedings and other publications, all of which are subject to rigorous technical review procedures.

The resource of some 38,000 practicing plastics engineers, scientists, and technologists has made SPE the largest organization of its type worldwide. Further information is available from the Society at 14 Fairfield Drive, Brookfield, Connecticut 06804, U.S.A.

Michael R. Cappelletti
Executive Director
Society of Plastics Engineers

Technical Volumes Committee
Claire Bluestein, Chairperson
Captan Associates, Inc.

Editor's Preface

This short volume is intended as a text and workbook for technicians employed in the plastics industry. The original German language edition was prepared at the behest of the government of the Federal Republic of Germany, the German Federal Association of Employers in the Chemical Industry and the German Chemical Workers Union. The book was put together and written at the Institut für Kunststoffverarbeitung (IKV) (Institute for Plastics Processing) at the Technical University of Aachen by Prof. Walter Michaeli with Helmut Greif, Hans Kaufmann and Franz-Josef Vossebürger.

Germany has long realized the necessity of organized post-high school teaching of trades to aspiring technicians. It is envisaged as the modern equivalent of the medieval apprenticeship program. Plastics processing is recognized as one of the modern trades deserving this treatment. The German program is described in some detail in Appendix 1. It is one of known success as witnessed by the success of Germany's many manufacturing firms. The skill of their well-trained technicians is proverbial.

In the United States, there has been increased interest in recent years in the post-high school training of technicians. However, little attention has been given to the system long established in Germany which could serve as a model.

It is hoped that this book will find use in educational programs for technicians taught both within industry and in schools/programs dedicated to this purpose.

James L. White
Institute of Polymer Engineering
University of Akron

Publisher's Note: Professor White edited the manuscript and adapted it for training courses of plastics technicians in the United States and in other countries where English is spoken. His effort is gratefully acknowledged.

How to Use this Book

This text and workbook provides an introduction to the world of plastics. Right from the start, the use of the plural "plastics" instead of the singular "plastic" indicates that we are dealing with a variety of materials that can differ markedly in processibility or in their reactions to the influence of heat.

Lessons

"Training for Plastics Technology" is divided into educational units that can be described as lessons, each lesson covering a distinct subject area. The lessons, approximately equal in length, are designed so that they can be arranged by each student in a meaningful educational sequence.

Key Questions

The key questions at the beginning of each lesson are intended to help the student approach the subject matter with certain questions in mind. The answers to these questions will become clear as the student works through the lesson.

Prerequisite Knowledge

It is not necessary to study the lessons in any particular sequence. Each lesson is therefore preceded by a list of other lessons or sections that are necessary to understanding the new material.

Subject Area

Since each lesson falls within a broader subject area, it is preceded by an indication of the subject area to which it belongs.

Review Questions

The review questions at the end of each lesson test the knowledge acquired. The student selects the correct answer from the choices provided and enters it in the blank space in the sentence. The answers can be checked against the lists of correct answers at the end of the book. If the chosen answer is incorrect, the student should review the corresponding subject matter.

Example: Compact Disk (CD)

In order to enhance the student's understanding of plastics and improve his or her ability to think in context, we have chosen a sample item formed from plastic which recurs in many of the lessons in the book. We use this product to show, for example, why a certain plastic is especially well suited to the manufacture of CDs. We also explore the question of whether this plastic can be recycled.

Appendices

The appendixes provide the interested reader with supplementary material pertaining to plastics. The reader can derive information on additional technical literature from the list of selected literature. The glossary is intended to contribute to a consistent understanding of the terms used in the book. It can be used as a type of brief lexicon. The profile of the German plastics processor provides more precise information about the training and the responsibilities associated with this plastics-related profession, as well as the various specialities and opportunities for continuing education and advancement within this career field.

Acknowledgments

The revised edition of this book has been a collaborative achievement. In addition to the authors, the following persons were involved: Thorsten Hamm and Selz Burak, who we would like to thank. The authors hope you will enjoy learning and working with this book.

Contents

Introduction

Key Questions

Where do we encounter plastics in our everyday lives?
How long have people used plastics?
What is a compact disk (CD) made of?

Contents

Plastics—Part of Our Everyday Lives

Plastics…

…are light

Plastics are so commonly used in our everyday surroundings that we take them for granted. We use freezer bags and plastic buckets in our households without wondering why these products are made of plastic. Why are buckets made of plastic instead of sheet metal or wood, as they used to be? In this example, weight plays the decisive role. The lighter plastic bucket is stable enough to carry water. So why use a heavier metal bucket?

…insulate against electric currents
…can be flexible

Why are electrical cables covered with plastic and not porcelain or a woven fabric? The plastic covering is more flexible than porcelain and more robust than a woven fabric. Yet it insulates the cable just as well, if not better.

…insulate against heat

Why is the inside of a refrigerator lined with plastic? For one thing, plastic is robust. It is also a poor conductor of heat, so it maintains low temperatures more effectively.

…can possess desirable optical properties

Why is a CD made of plastic? Polycarbonate (PC) is a plastic as transparent as glass, yet lighter and not as fragile.

…are inexpensive materials

Of course, we must also consider the price in all of these examples. The use of plastics is often the most inexpensive technical solution, especially for mass-produced goods. Later we'll consider why this is so and what problems are often overlooked in choosing this solution (e.g., waste disposal).

Plastics—Multifaceted Materials

wood

natural rubber

Before plastics became known, lightweight materials could only be obtained from nature. Wood is easy to process and is strong and flexible. It can also be permanently shaped by special processes. Natural rubber, is elastic and stretchable.

natural materials

All technical problems cannot be solved with the properties of natural materials, however. So a search for new materials with the required properties began. Not until the twentieth century did chemists learn enough about the molecular structure of natural materials (e.g., natural rubber) to make it possible to produce these materials synthetically.

ideal properties

Many of the plastics manufactured today exhibit properties that are far superior to those of natural materials. We now possess materials with properties that are ideally matched to the respective application and serve the widest conceivable variety of purposes.

material properties

It is impossible to determine the purpose for which a plastic article is best suited merely by observing its external appearance. We also need to know something about the internal structure of the material. This gives us information about its density, conductivity, permeability, or solubility, for example—"material-specific properties."

Plastics—Young Materials

plastics model

The purposeful conversion of natural materials into the materials known today as "plastics" first began in the nineteenth century. But this activity did not acquire scientific significance until the 1920s, when the German chemist Professor Hermann Staudinger developed his model of the structure of plastics. In 1953, Professor Staudinger (1881–1965) received the Nobel Prize for this research.

Nobel Prize

worldwide boom

The worldwide boom in the plastics industry began after World War II. Coal was first used as the basic material, but a switch to petroleum took place in the mid-1940s in the USA and in the mid 1950s in Europe, which allowed previously worthless refining constituents (which occurred as separation products in the process of cracking crude oil) to be put to effective use. The rapid increase in plastics production experienced its first moderate setback during the oil crisis of 1973. These materials have nonetheless recorded an above-average, dynamic rate of development through the present day.

substitution of classic materials

However, plastics can only be used with optimal effectiveness when their special characteristics are taken into account. In the substitution of plastics for classic materials like wood or metal, it is especially important to create designs with plastics in mind in order to fully exploit the variety of possibilities they present. It is important to be familiar with the appropriate processing methods and the respective characteristic material values.

compact disk (CD)

This plastic-based approach requires a fundamental understanding of manufacturing and processing methods, as well as material characteristics. This book is intended to provide a fundamental, yet comprehensive overview of the subject of plastics. We also intend to follow a modern plastic article, a compact disk (CD), from its beginnings as crude oil until it becomes garbage. The CD is a high-tech product especially well suited as an example of modern plastics processing.

Lesson 1

Plastics Fundamentals

Key Questions

How can plastics be defined?
What are plastics made of?
How are plastics classified?
What is the chemical structure of a simple plastic?
What plastic is a CD made of?
Are plastics recyclable?
What are the properties of plastics?
Where are plastics used?

Subject Area

Plastics Fundamentals

Contents

1.1 What Are "Plastics"?

generic term

The name "plastic" does not just stand for a single material. Just as the word "metal" does not signify only iron or aluminum, the name "plastic" is a generic term for numerous materials that differ in structure, properties, and composition. The properties of plastics vary so widely that they are often used to replace or supplement conventional materials, such as wood or metal.

macromolecules

All plastics have one thing in common, however. They are the result of the tangling or linking of very long molecular chains (chain molecules) known as "macromolecules" (macro = large). These macromolecules often consist of more than 10,000 individual structural elements. The individual elements of these molecular chains are arranged one after the other like beads on a necklace. We can imagine a plastic as something similar to a ball of wool consisting of many individual threads. It is very difficult to pull a single thread from the ball. This is similar to a plastic in which the macromolecules hold one another fast. The many individual structural elements that make up the macromolecules are called "monomer" units (mono = single, meros = part). Therefore, the macromolecules, and thus the plastics themselves, are also generally called "polymers" (poly = many).

definition

Plastics are materials whose essential components consist of macromolecular organic compounds. These compounds are created synthetically or by the conversion of natural products. As a rule, these materials can be shaped or undergo plastic deformation when processed under certain conditions (e.g., heat or pressure).

1.2 What Are Plastics Made of?

monomers

The basic substances for polymers are called "monomers." It is often possible to produce several different polymers from the same individual basic substances by altering the production process or creating different mixtures.

raw materials

The basic materials for the monomers are primarily crude oil and natural gas. Because carbon is the only element essential for polymers, it is theoretically possible to create monomers from wood, coal, or even atmospheric CO_2. But these substances are not used because it is inexpensive to produce monomers from gas and oil.

refinery products

Some monomers have occurred for many years as waste products from the production of gasoline and heating oil. The great consumption of plastics now makes it necessary to intentionally produce these "waste monomers" in refineries.

1.3 How Are Plastics Classified?

Three overall groups of plastic materials are distinguished from one another. Figure 1-1 presents each of these groups with examples.

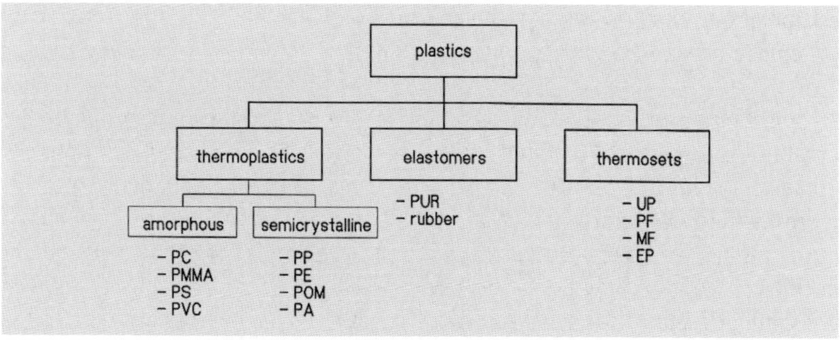

Figure 1-1 Classification of plastics

thermoplastics

Thermoplastics (thermos = warm; plasso = shape, shapable) are fusible and soluble. They can be melted repeatedly and are capable of dissolving or at least swelling in many solvents. They vary from a soft condition to a ductile or rigid hardness at room temperature. Amorphous (amorph = formless) thermoplastics are similar to glass with respect to molecular structure and transparency. They are distinguished from semicrystalline thermoplastics, which have a milky, opaque appearance. When a plastic is as transparent as glass, it is reasonably safe to assume that it is an amorphous thermoplastic (though there are exceptions, e.g. polymethylpentene). Thermoplastics make up the largest proportion of plastics with respect to quantity.

amorphous thermoplastic

semicrystalline thermoplastics

CD

We therefore want to manufacture the cover of our CD case from an amorphous thermoplastic. After all, it should be transparent, so that we can read the list of titles through it. The CD itself is also made of a transparent plastic. In most cases, it is initially coated with aluminum on one side. This is accomplished by a vacuum deposition process. (The aluminum works like a mirror.) The CD is then printed. As a result of this treatment, the laser beam does not pass through the CD, but is reflected.

thermosets

Thermosets are hard plastics, tightly cross-linked, i.e. with the cross-links close together, impervious to plastic deformation, infusible, and highly temperature resistant. Because thermosets are very densely cross-linked, they are impossible to dissolve and very difficult to swell. Electrical outlets are manufactured from thermosets.

elastomers

Elastomers (elastic = springy; meros = part) cannot be melted or dissolved, but they can be swelled. They are lightly cross-linked, i.e. with the cross-links far apart, and therefore exist in a soft, elastic condition at room temperature. Examples of elastomers include tires and rubber seals for canning jars.

In some types of elastomers, called “thermoplastic elastomers”, the cross-links disappear on heating. These materials can readily be reprocessed unlike conventional elastomers.

1.4 How Are Plastics Identified?

Standard practice provides for each plastic to be identified by a sequence of symbols (an acronym) that indicates its chemical structure. Additional letters (codes) identify its use, fillers, and basic properties, such as density or viscosity. An example is given in Figure 1-2.

Identification of the Plastic

PE - HD

Material Name

linear high-density polyethylene

Abbreviation for the Basic Polymer Product

PE = polyethylene

Code Letters for Additional Properties

H = first code letter for special properties: H = high

D = second code letter for special properties: D = density

Figure 1-2 Example of standard plastic identification

quantities and values

For the moment, we intend only to introduce the numerous quantities and values presented here. You may want to review this section after reading the rest of the book. This may help you put many of the unfamiliar concepts (such as "molding compound" or "melt flow range") into proper perspective.

1.5 What Are the Physical Properties of Plastics?

Plastics Are Light

lightweight construction materials

Plastics are suitable lightweight construction materials, usually lighter than metals or ceramics. Many plastics are lighter than water and can therefore float. They are used for lightweight components in aircraft construction, automobile production, packaging, and sporting goods. For example, aluminum is three times as heavy as polyethylene (PE), and steel is eight times as heavy as polyethylene.

CD

The CD turns at a speed of 200 to 500 revolutions per minute. In order for the motor of the CD player to start quickly and yet remain small, it is important that the CD be light.

Plastics Are Easy to Process

processing temperature

The processing temperatures of plastics range from room temperature up to approximately 250°C (508°F)—in special cases even up to 400°C (778°F). (Compare the low temperature to a processing temperature of more than 1400°C (2578°F) for steel!) The low temperature makes processing less difficult and requires relatively little energy. This is why production costs are rather low, even for complicated parts. The various processing methods, such as injection molding or extrusion, will be presented later in detail.

Properties of Plastics Can Be Purposefully Optimized

additives

The low processing temperatures required for plastics also makes it possible to incorporate additives of various types, such as dyestuffs and pigments, fillers (e.g., wood flour or mineral powder), reinforcements (e.g., glass or carbon fibers), and blowing agents for producing foamed plastics.

colorants

Colorants make it possible to color the material. In most cases, this makes a subsequent lamination process unnecessary.

fillers

Inorganic fillers in the form of a powder or sand can be used in large proportions (up to 50%). They increase the modulus of elasticity and compressive strength and help make the plastic more cost effective. Organic fillers such as woven fabrics of (textile) fibers or cellulose webs increase toughness. Carbon black is incorporated in automobile tires (elastomers!), for example. It improves the mechanical properties (abrasion resistance) and increases thermal conductivity and resistance to light. The incorporation of plasticizers (certain esters and waxes) can alter the mechanical characteristics of a hard plastic to such a degree that it reaches an elastomer-like condition.

reinforcements

Fibers of glass, carbon, and aramid (e.g. Kevlar) are examples of materials used as reinforcements. They are used in various forms, as short or long fibers, as a weave or a mat. Effective embedding of fibers in the plastic allows a great increase in strength and rigidity.

blowing agents

Blowing agents are used to create synthetic foamed plastics, whose density can be reduced to one hundredth of the density of the starting material. Foamed plastics possess especially favorable thermal insulating properties and make it possible to manufacture very light construction components.

Plastics Possess Low Conductivity

insulation

Plastics insulate against electric current (as in electrical cables) and against cold or heat. Examples include refrigerators and plastic cups. The ability of plastics to conduct heat is approximately 1000 times less than the thermal conductivity of metals.

electrical conductivity

The reason for the reduced electrical conductivity of plastics in comparison to metals is that plastics possess practically no free electrons. These electrons are responsible for carrying heat and electrical current in metals. This property of plastics can be modified to a considerable degree by the use of additives.

thermal conductivity

Plastics are suitable for use as insulating materials. Their low degree of thermal conductivity leads to problems in processing, however. For one thing, the heat of melting is conveyed very slowly into the inner regions of the material.

As a result of their favorable insulating effect, plastics can acquire an electrostatic charge. If conductive materials (e.g., metal powder) are mixed with the plastic before processing, the insulating effect is reduced. This in turn reduces the tendency to static charging.

Plastics Resist Many Chemicals

corrosion

CD

The mechanism that bonds the atoms in plastics is very different from the corresponding mechanism in metals. For this reason, plastics are not as vulnerable to corrosion as are metals. Some plastics are very resistant to acids, bases, or aqueous salt solutions. In many cases, however, they are soluble in organic solvents, such as gasoline or alcohol. Therefore, the CD should not be cleaned with turpentine when it becomes dirty, because this agent could damage the plastic.

solvents

It is best to dissolve a plastic in a solvent of chemical composition similar to the composition of the plastic. As the expression goes, "like dissolves like."

Plastics Are Permeable

diffusion

material values

The penetration of one substance (for example, a gas) through another is called "diffusion." A high degree of permeability to gases—resulting from large distances between molecules (i.e., low density)—is sometimes a disadvantage. This permeability differs from one plastic to another, however. But this permeability is practical in some applications, such as membranes for seawater desalination plants, certain packaging films, or organ replacements. In order to find suitable plastics for a given field of application, researchers obtain the density or other characteristic material values from manufacturers' specifications or data sheets, for example.

Plastics Are Recyclable

recycling

incineration

Plastics can be reutilized or recycled by various methods. It is also possible to obtain energy from the incineration of various plastics that cannot be recycled economically. However, some materials cannot be incinerated without problems and need specific incineration and special filter technique. This especially applies to plastics containing chlorine (such as PVC) or fluorine (such as PTFE, better known by the trade name Teflon). The incineration of these plastics creates poisonous gases.

product identification

Today, it is obligatory to provide each plastic product with an identification symbol. When the product is recycled, this symbol makes it possible to determine which plastic it is made of. It is thus possible to recycle waste according to its different types.

Other Properties of Plastics

flexibility

Some plastics are flexible. Although strength and modulus of elasticity vary widely between plastics, they usually lie considerably lower than the corresponding properties of metals. In many cases, their high degree of flexibility is an advantage in manufacturing and application.

impact resistance

Several plastics display a better impact resistance than mineral-based glass without sacrificing any of the favorable optical properties of glass. In other words, plastics are not as fragile as glass (although they do scratch more easily). For this reason, plastics are becoming increasingly common as a replacement for glass in applications such as construction, automobile production, and eyeglasses.

Review Questions

No.	*Question*	*Choices*
1	Plastics are categorized in the following groups: thermoplastics, elastomers, and ______________________.	monomers thermosets
2	Thermoplastics are categorized in the following two subgroups: amorphous thermoplastics and ______________________ thermoplastics.	thermosetting semicrystalline
3	Thermoplastics ______________________ on heating.	melt do not melt
4	Thermosets are densely cross-linked and therefore _____________.	soluble insoluble
5	Elastomers are ______________________ cross-linked.	densely loosely
6	Elastomers are ______________________________________.	swellable not swellable
7	Thermoplastic elastomers behave like ______________________ on heating.	elastomers thermoplastics
8	Most plastics are ______________________than metals.	lighter heavier
9	Plastics have a ______________________ processing temperature than metals.	higher lower
10	Different plastics display ______________________ degrees of permeability to gases.	identical different
11	Plastics are very ______________________ insulators against heat and electrical current.	poor good
12	Most plastics ______________________ be recycled.	can cannot

Lesson 2

Raw Materials and Polymer Synthesis

Key Questions

What raw materials are plastics made of?
What are the steps in refining petroleum to obtain the basic substances used in making plastics?
How are plastics structured?
What is a monomer?
What are macromolecules and chain elements?
What methods of polymer synthesis exist?

Subject Area

Chemistry of Plastics

Contents

Prerequisite Knowledge

Plastics Fundamentals (Lesson 1)

2.1 Raw Materials for Plastics

chemistry of plastics

The raw materials used in the production of plastics are natural substances, such as cellulose, coal, petroleum, and natural gas. The molecules of all these raw materials contain carbon (C) and hydrogen (H). Oxygen (O), nitrogen (N), and sulfur (S) can also be involved. The most important raw material for plastics is petroleum.

petroleum

Figure 2-1 shows the proportions of the various products made from petroleum as percentages of total production. It is apparent that only 4% of all petroleum is processed into plastics.

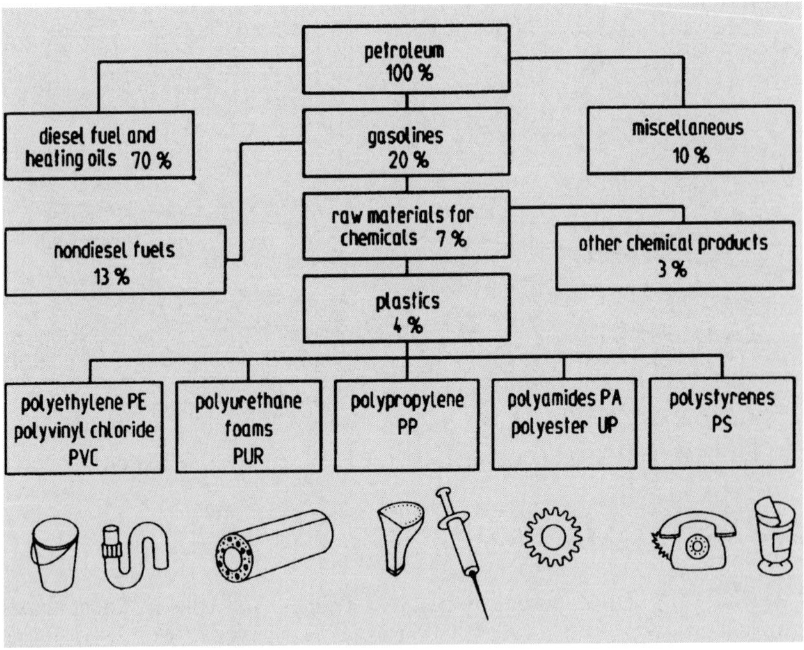

Figure 2-1 Breakdown of products made from raw materials (Note: UP = unsaturated polyesters; other polyesters, such as polyethylene terephthalate (PET) are also important).

intermediate steps

distillation

Plastics are not made directly from petroleum, however. Several intermediate steps are required. Petroleum is separated into its different components by distillation. This process for separating liquids occurs in a refinery. Separation is accomplished by taking advantage of the boiling points of these different components. Separation produces gas for municipal use, gasoline, kerosene, and fuel oil. Bitumen occurs as a residue from the distillation process and is used in road construction.

cracking

The most important distillate for plastics production is crude gasoline. A thermal breakdown process divides the distilled gasoline into ethylene, propylene, butylene, and other hydrocarbons. This process is also called "cracking." By varying the process temperature, it is possible to control the proportions in which the individual separation products are obtained. For example, more than 30% ethylene is obtained at 850°C (1588°F).

basic substances

Other products (e.g., styrene and vinyl chloride) can then be obtained from ethylene in subsequent reaction steps. In addition to these two substances, ethylene, propylene, and butylene are also basic substances (monomers) from which plastics are manufactured.

It is a known fact that all work processes need certain energy (pressure, heat, motor energy, etc.). Figure 2-2 compares various production methods of the petrochemical industry and how they save energy. One can see how much energy is required (measured in tons mineral oil) to manufacture various products.

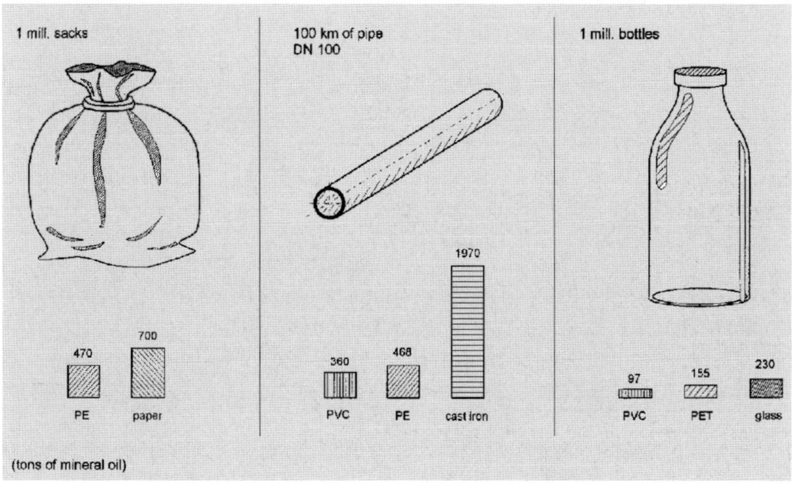

Figure 2-2 Comparison of required energy to produce various products.

monomers
macromolecules

2.2 Monomers and Polymers

The basic substances used in making plastics are called "monomers" (mono = single, meros = part). The macromolecules of plastics are manufactured from these basic substances. The term "macromolecule" is derived from the large size of these molecules (macro = large). These large molecules result from the combination of many thousands of monomer molecules.

polymer

Before the macromolecule is formed, the monomers exist separately (see Figure 2-3). The molecule composed of many of these monomers is called a "polymer" (poly = many). A chemical reaction is required to convert the individual molecules to a macromolecule.

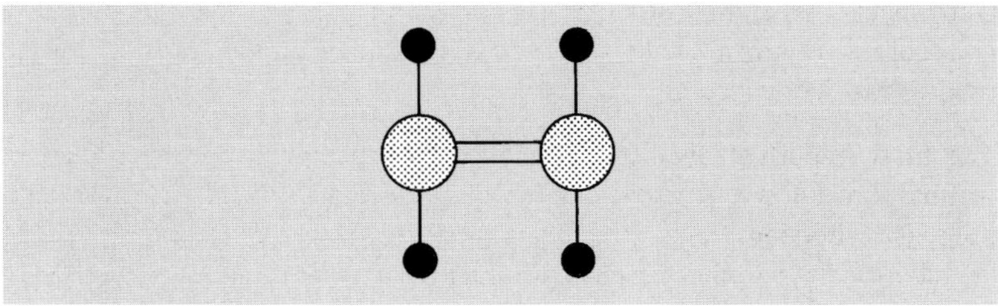

chain elements

Figure 2-3 Monomer molecule (schematic)

Because the macromolecules are formed from many identical monomers in the simplest case, they consist of a sequence of identical chain elements (see Figure 2-4). Each chain molecule has a continuous line of chain elements.

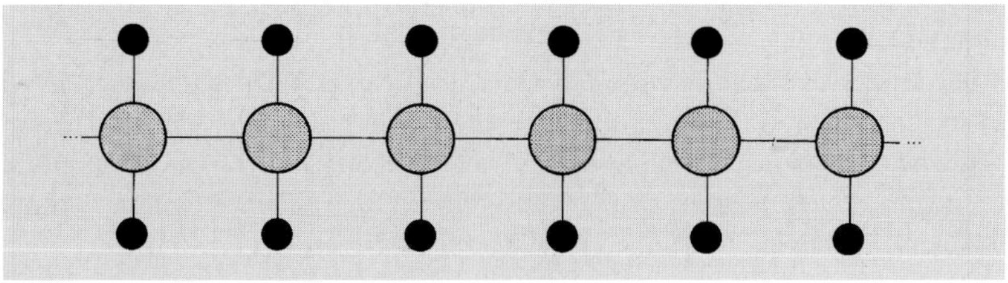

backbone

Figure 2-4 Macromolecule (chain elements)

Other elements are attached to this line. The continuous line of the macromolecule—the "backbone"—is often composed of nothing but the element carbon (C). Oxygen (O) and/or nitrogen (N) is also present in some cases. Carbon atoms have the property of readily forming chains with oxygen, nitrogen, or other carbon atoms. This property is not as prominent in other chemical elements.

side chains

Various other elements or element groups are attached to the backbone. One example is hydrogen (H) (see Figure 2-4). If these element groups consist of chain elements that also form the actual molecule chain, they are known as "branches" or "side chains." These branches occur to some degree in most plastics.

polyethylene

2.3 Polyethylene Synthesis

One example of a macromolecular substance is polyethylene (see Figure 2-5).

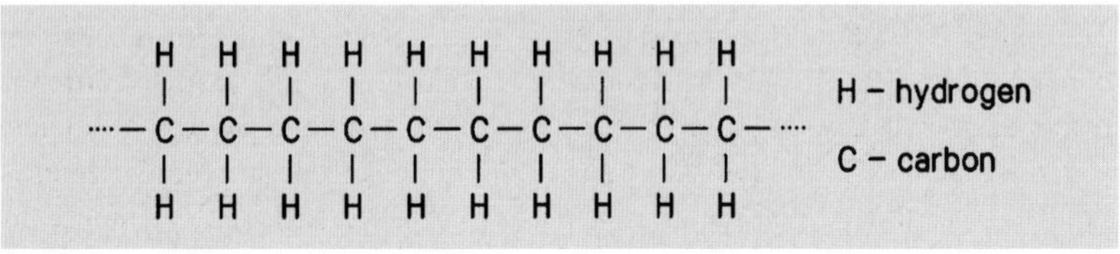

Figure 2-5 Structure of a polyethylene string molecule

The monomer from which polyethylene is made is called ethylene. It consists of only carbon and hydrogen, as shown by the structural formula in Figure 2-6.

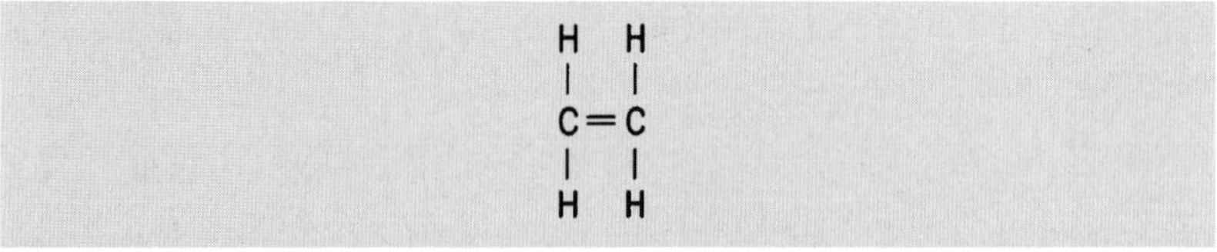

Figure 2-6 Structural formula for ethylene (monomer of polyethylene)

bonding

double bond

The lines on the figure represent the bonds between the atoms. A bond consists of a pair of electrons shared by two atoms. The double lines between the carbon atoms represent a double bond. The double bond is important in the reaction that forms the macromolecule. The ethylene molecules are activated one after the other and gradually form a macromolecule. The structural formula for this macromolecule is shown in Figure 2-7.

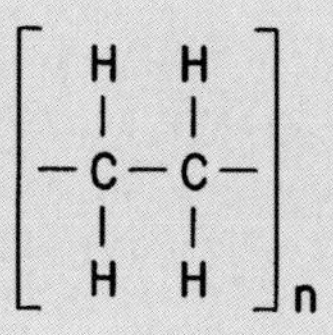

Figure 2-7 Structural formula for polyethylene (PE)

tangling

The letter "n" is a number that indicates how often this unit is repeated within the macromolecule, usually greater than 10,000. Because the macromolecules are not formed one at a time, but simultaneously, they become tangled (see Figure 2-8). The resulting material is a plastic.

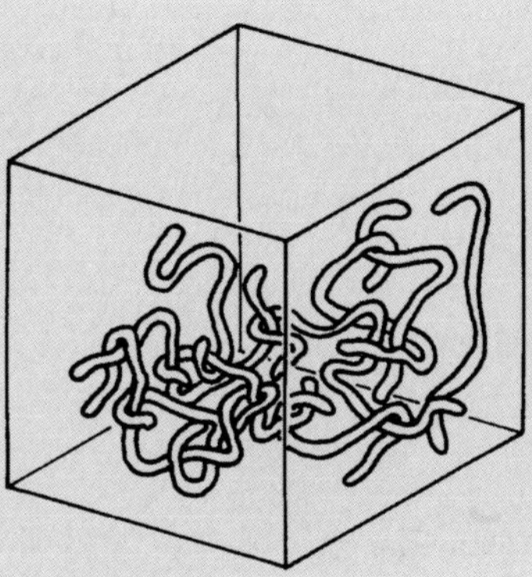

Figure 2-8 Tangled macromolecules

process

The macromolecules are formed by different reactions, as determined by the type of monomer. Plastics are manufactured by three fundamentally different reactions or processes. These processes are called synthetic processes, because a new material is being synthesized (i.e., combined) from structural elements. In this case, the new material is a plastic, and the structural elements are monomers. The technical names for plastics manufactured by these processes are based on the names of the respective processes (see Figure 2-9).

synthetic process	**product name**	**examples/abbreviations**
addition (or chain-reaction) polymerization	addition polymers	polypropylene / PP polyethylene / PE
polyaddition (or step addition)	polyadducts or polyaddition products	polyurethane / PUR epoxy resins / EP
polycondensation (or step polymerization)	condensation polymers	polycarbonate / PC polyamide / PA

Figure 2-9 Product names and examples of plastics

Review Questions

No.	*Question*	*Choices*
1	The raw materials used in producing plastics include coal, petroleum, and ________________________.	steel aluminum natural gas
2	Monomers for making plastics are obtained from petroleum by the processing steps of distillation and ____________________________.	conversion cracking
3	Ethylene, vinyl chloride, and ____________________ are examples of monomers produced from petroleum and used in making plastics.	kerosene propylene
4	A macromolecule can be pictured as a long ____________________.	chain tube
5	"Monomer" is the general chemical name for the molecules from which plastic is manufactured. The plastic composed of many such molecules is therefore called a ______________________________.	copolymer polymer thermoset
6	In most cases, the continuous line (back-bone) of the macromolecule is formed by the chemical element ___________________________.	carbon (C) oxygen (O) nitrogen (N)
7	The abbreviations for plastics often begin with the letter "P," which stands for ________________________.	partial poly particle
8	After the macromolecules have been formed, they exist in a ________________________ condition.	tangled stretched
9	Polyethylene (PE) has a very simple structure. It is composed of only the elements hydrogen (H) and _________________________.	carbon (C) fluorine (F) oxygen (O)

Lesson 3

Methods of Polymer Synthesis

Key Questions

What are the special features of addition polymerization, and how does the reaction proceed?
What is the difference between homopolymerization and copolymerization?
What are the special features of polycondensation, and how does the reaction proceed?
What are the special features of polyaddition (step addition), and how does the reaction proceed?
What are examples of plastics manufactured by addition polymerization, polycondensation, and polyaddition (step addition)?

Subject Area

Plastics Chemistry

Contents

Prerequisite Knowledge

Raw Materials and Polymer Synthesis (Lesson 2)

3.1 Addition Polymerization

A Double Bond in the Monomer

double bond

The double bond existing between two carbon atoms in the monomer plays a decisive role in addition polymerization. We will show this in the example of vinyl chloride (Figure 3-1).

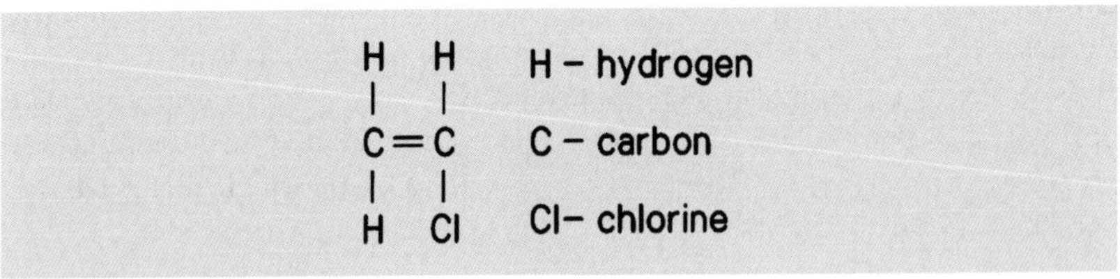

Figure 3-1 Structural formula for vinyl chloride

double bond splitting

In vinyl chloride, as elsewhere, each bond within the molecule consists of two electrons. The double bond therefore consists of two bonds, each of which joins two electrons. One of these two bonds can be split rather easily. In other words, it can be broken down to produce two (unpaired) electrons.

Formation of the Macromolecule

macromolecules

radical

Bond splitting leads to the formation of the molecular chain i.e. the macromolecule. It begins with the splitting of the double bond, which is caused by another particle, a radical, for example. Radicals are elements or element groups with a high degree of reactivity. This means that they react quite readily with other molecules. The reason for this is that each radical has a free individual electron that readily forms a bond with another electron. Figure 3-2 shows the splitting of the double bond in vinyl chloride by a radical (R).

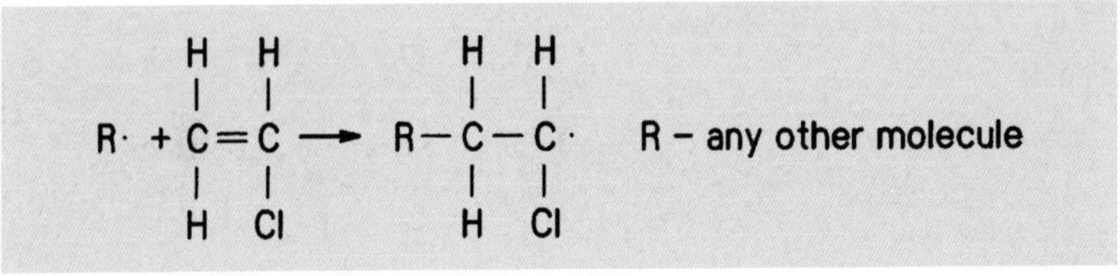

Figure 3-2 Splitting of double bond

chain formation

When the bond is split, the electron of the radical forms a new bond with one electron from the split bond. The other electron from the split bond is now located on the opposite side of the vinyl chloride molecule. The side with this free electron can now split other double bonds, in turn. A long chain thus grows from these beginnings (see Figure 3-3).

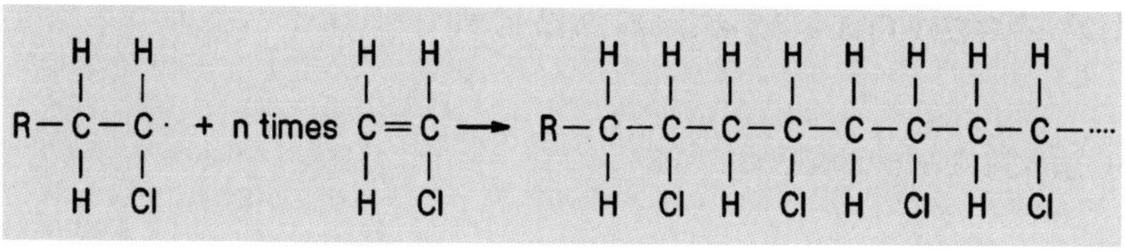

Figure 3-3 Chain formation

length of chain elements

Growth ends when one end of a chain meets another chain end or a radical. Because there are initially many more vinyl chloride monomers than chain ends or radicals, however, the chains become very long before they stop growing. The length of these chains is very important in determining the properties of the plastic. The length is specified as n, which represents the number of recurrent chain elements (see Figure 3-4).

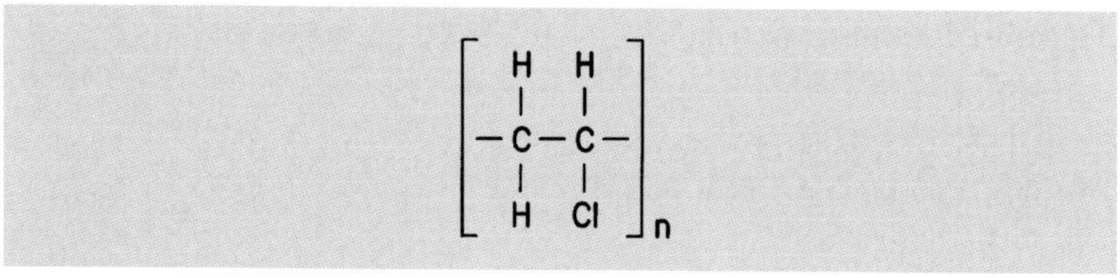

Figure 3-4 Repeat unit

The number *n* is often greater than 10,000. In order to visualize the possible length of such a molecule, imagine that it were enlarged one million times. It would then be 20 cm (8 in) thick and 1 km (3300 ft) long.

polymers

The products of addition polymerization are called polymers (see Figure 3-5).

polymers	**abbreviation**	**products**
polyethylene	PE	protective films, packaging films, bottles, pipes, transport containers, electrical accessories, coverings, fittings, construction of chemical apparatus
polypropylene	PP	instrument housings, washing machine parts, electrical installations, pipes, fittings, construction of apparatus
polymethyl methacrylate	PMMA	glazing, taillights, sanitation components, signs, lenses, drawing instruments, skylights

Figure 3-5 Polymers and their applications

catchword "coupling"

How can we firmly impress the concept of the polymerization process upon our minds? The cars of a train can only be coupled together when coupling devices are provided at the front and rear of each car. A macromolecule chain is formed in a similar way in the polymerization process. The chain results from the coupling of the individual monomers by means of the electrons from the split double bonds. So the catchword for polymerization is "coupling."

Copolymerization

copolymer

One or more types of monomers can be used simultaneously to manufacture a plastic by addition polymerization. If only one monomer is used in polymerization, a homopolymer results. The manufacture of a polymer from two or more different monomers is called "copolymerization" (co = with, together). The result is a copolymer. The arrangements of the different monomer structural elements in the copolymer can vary. The properties of the plastic can be influenced by the choice of different monomers in copolymerization.

3.2 Polycondensation

polycondensation

Small molecules—usually water molecules—are typically split off in the polycondensation reaction. In organic chemistry, this process is called "condensation," hence the name of this type of plastics manufacturing. The chemical formula for water is H_2O. A water molecule is thus composed of two hydrogen atoms (H) and one oxygen atom (O).

functional groups

The formation of macromolecules by the polycondensation reaction requires molecules that possess two or more "functional groups." (see Figure 3-6).

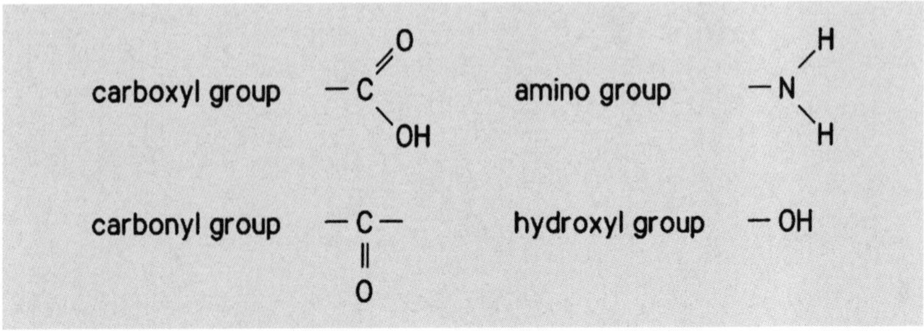

Figure 3-6 Functional groups

However, a bond between two molecules can only be formed in the presence of two different functional groups from which the elements split off and then "condense" as water or some other small molecule.

types of molecules

In order for the polycondensation reaction to form a continuous chain, we must have the following types of molecules: either one type of molecule with at least two different functional groups or at least two different molecule types, each of which features two or more similar functional groups.

One example of condensation is the reaction in which an amide is created from two molecules. The plastic made by polycondensation of many molecules is therefore called a "polyamide," or a "nylon"—after an early DuPont trademark. One example of polycondensation is the reaction between hexamethylene diamine and adipic acid, which results in polyamide 66 (PA66), also called nylon 66 (see Figure 3-7).

polyamide

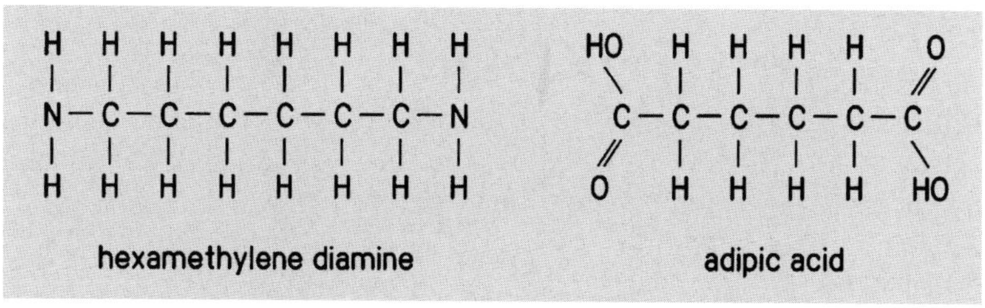

Figure 3-7 Structural formulas

The reaction happens in two steps. During the first step, the particles split from the functional groups. During the second, the macromolecule polyamide and water is formed (see Figure 3-8.) The reaction can also happen in reverse. Therefore, the water has to be constantly removed during the production of polyamide.

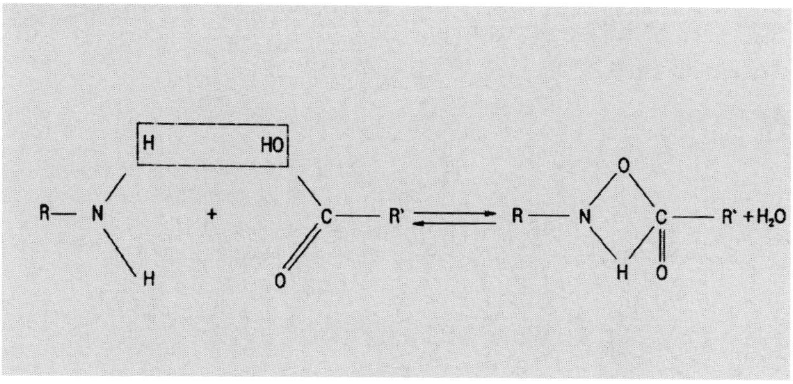

Figure 3-8 Reactions to form polyamide

The molecules separated in the reaction (water in this example) must be constantly removed during polycondensation in order to allow the reaction to continue and form very long chains. The polycondensation reaction has no termination step. Plastics resulting from polycondensation are called "condensation polymers" (see Figure 3-9).

water removal

condensation polymers

condensation polymers	abbreviation	examples
phenol-formaldehyde (resin)	PF	handles for shift levers, switch parts, automobile ashtrays, heaters, flatirons, pots and pans, lamp sockets
unsaturated polyesters	UP	glass-fiber reinforced in construction of boats and vehicles, device housings
polycarbonate	PC	housings for office machinery and household appliances, observation glasses, CDs, camera housings, signal lamps
polyamides	PA	gears, rollers, housings for electrical devices, anchor pegs

Figure 3-9 Condensation polymers and applications

CD

Our CD also consists of a plastic manufactured by polycondensation, namely polycarbonate (PC).

catchword "splitting off"

How can we firmly impress the concept of the polycondensation process upon our minds? Water is split off and removed in the polycondensation process. The catchword for polycondensation is therefore "splitting off".

3.3 Polyaddition (step addition)

polyaddition

The reaction of polyaddition proceeds similar to the polycondensation reaction. The difference is that in polyaddition no elements are separated and no small molecules are split off. Instead, a hydrogen atom migrates from one functional group to another.

Two different functional groups are needed to form a bond, just as in polycondensation. Each of the monomers must again possess at least two functional groups. The macromolecule is again formed by using one type of molecule with at least two different functional groups or at least two types of molecules, each having two or more similar groups.

reaction sequence

The reaction can be represented in three steps.

Step 1: A hydrogen atom that can be easily separated is present at one molecule end, and a bond that can be easily split is present at another molecule end.

Step 2: The hydrogen atom is separated, and the bond of the other functional group is split.

Step 3: The hydrogen atom forms a new bond with one of the electrons of the split bond. The other electron from the split bond forms a new bond at the location from which the hydrogen atom has separated, and the chain is thereby lengthened.

Figure 3-10 shows a schematic representation of the polyaddition reaction.

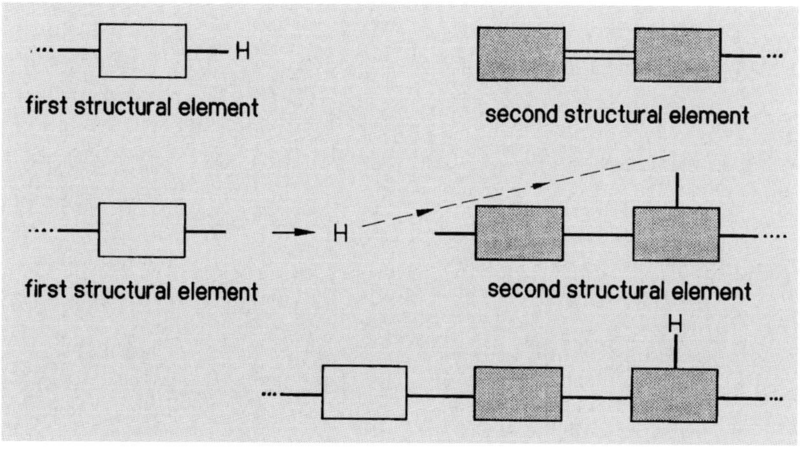

Figure 3-10 Polyaddition reaction

Polymers created by the polyaddition reaction are called "polyaddition products." A few such polymers and their applications are presented in Figure 3-11.

polyaddition products

polyaddition products	abbreviation	products
polyurethane	PUR	heels, rolls, bearings, clutch plates
polyurethane foams	PUR foam	insulating and upholstering foams for furniture, buildings, clothing
epoxies	EP	adhesives, container coatings, tools (with fiber reinforcement)

Figure 3-11 Addition polymers and their applications

The characteristic feature of this chemical reaction is the exchange of an atom, which changes from the functional group of one reaction partner to the functional group of the other reaction partner.

catchword "exchange"

NORTHEAST WISCONSIN TECHNICAL COLLEGE
LIBRARY - 2740 W MASON ST
GREEN BAY, WI 54307

Review Questions

No.	*Question*	*Choices*
1	The ______________________ bond plays the decisive role in addition polymerization.	single double
2	The catchword for addition polymerization is _________________.	"coupling" "exchange" "water removal"
3	Polymers manufactured from different monomers are called ______________________.	homopolymers copolymers
4	Two examples of plastics manufactured by addition polymerization are polyethylene (PE) and ______________________.	polycarbonate (PC) polypropylene (PP)
5	In organic chemistry, condensation means the _________________ of elements from the reactants to form a small molecule.	splitting off evaporation
6	Polycondensation usually involves the removal of ______________________.	hydrogen carbon dioxide water
7	The molecules used in polycondensation must possess ___________ functional group(s) so that macromolecules can be formed.	two or more one no
8	Examples of plastics manufactured by polycondensation include phenol-formaldehyde (PF) and ______________________.	polyethylene (PE) polycarbonate (PC)
9	The catchword for polycondensation is ____________________.	"exchange" "splitting off" "coupling"
10	Just as in polycondensation, the monomers used in polyaddition must possess two or more ______________________.	functional groups hydrogen atoms
11	Examples of plastics manufactured by polyaddition include polyurethane (PUR) and ______________________.	epoxies (EP) polyamides (PA)
12	The catchword for polyaddition is ______________________.	"coupling" "splitting off" "exchange"

Lesson 4

Bonding Forces in Polymers

Key Questions

What types of forces exist within a polymer?
What are the differences between these forces?
How does temperature affect these forces?

Subject Area

Physics of Plastics

Contents

Prerequisite Knowledge

Raw Materials and Polymer Synthesis (Lesson 2)

4.1 Bonding Forces within Molecules

atomic bonds

The atoms of the monomer molecules that compose the macromolecules are connected by atomic bonds (also known as "covalent bonds"). We can think of these bonds as forces that hold two atoms together. Graphic representations of molecules generally indicate these bonds with lines. One example is the monomer ethylene (Figure 4-1).

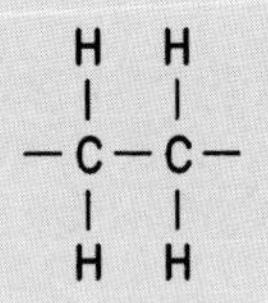

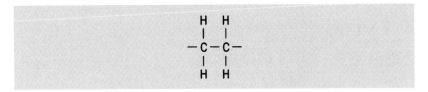

Figure 4-1 Structural formula for ethylene

number of bonds

A distinction between single, double, and triple bonds is based on the number of bonds between two atoms. As shown in the figure, ethylene contains a double bond between the two carbon atoms (C) and a single bond between each hydrogen atom (H) and a carbon atom (C). The double bond is an unsaturated bond. "Unsaturated" means that the bond can easily be split. This makes it possible to create another bond with other atoms. These bonding forces also occur in the macromolecules of the plastic.

4.2 Intermolecular Forces

intermolecular forces

Forces exist not only within a molecule (intramolecular forces) but also between adjacent molecules. The latter forces are thus called "intermolecular." They cause two molecules to attract one another with a certain amount of force so that they cannot move apart by their own energy) (see Figure 4-2).

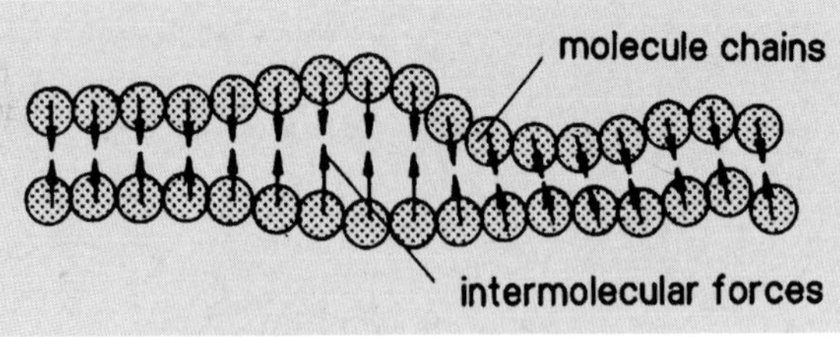

Figure 4-2 Intermolecular forces

strength

These intermolecular forces also operate between the tangled macromolecules within a plastic. They provide the plastic with most of its strength, since they hold the molecules together and prevent them from easily "slipping" away from one another. The bonds can be pictured as the tiny hooks between the two strips of fabric of a velcro fastener. The strips of fabric hold tightly to one another by means of these hooks. They can only be separated by a strong pull.

The intermolecular bonds are not as strong as the atomic (covalent) bonds, however. When a load is applied, the bonds between the molecules are the first to be broken. We can observe this with another look at the velcro fastener. Imagine that the atomic bonds are the forces within each strip of woven fabric holding it together. When we pull forcefully on the ends of the fastener, the strip of fabric does not tear first. Instead, the hooks separate and the strips of fabric slip away from one another. The intermolecular bonds are the first to separate.

atomic bonds

4.3 Effects of Temperature

heat

Heat is manifested in the movement of molecules. As the temperature increases, the molecules move more vigorously. This movement reduces the intermolecular forces. Above a certain temperature, these forces are completely eliminated, allowing free movement of the molecules that previously had been connected by these forces. If the temperature is then reduced, the movement of the molecules is reduced in turn, and the forces develop again.

thermal motion

The covalent bonds between the atoms of a molecule are not separated by thermal motion. They are much stronger and are not destroyed until much higher temperatures are attained. In contrast to intermolecular forces, the atomic forces do not develop again when the temperature falls. The molecule remains destroyed.

thermal expansion

The increasing movement of the molecules also results in a greater space requirement. The plastic therefore expands as its temperature increases. This change in volume with the change in temperature—"thermal expansion"—varies in intensity from one material to another. Different plastics also exhibit different degrees of thermal expansion. One measure of the change in length is the coefficient of linear thermal expansion. The higher this value, the more the material expands when heated) (see Figure 4-3).

material	abbreviation	coefficient of linear thermal expansion α [1K × 10^{-6}] at 50°C (122°F)
polyethylene	PE	150–200
polycarbonate	PC	60–70
steel	St	2–17
aluminum	Al	23

Figure 4-3 Coefficient of thermal expansion of various materials

Review Questions

No.	*Question*	*Choices*
1	The bonds between the atoms within a macromolecule are called ________________________ or covalent bonds.	intermolecular bonds atomic bonds
2	The bonds operating between two macromolecules are called ________________________.	intermolecular bonds atomic bonds
3	The forces of an atomic bond are considerably __________________ than those of an intermolecular bond.	stronger weaker

Lesson 5

Classification of Plastics

Key Questions

What are the categories into which plastics are divided?
What are the criteria for classifying them?

Subject Area

Plastics Fundamentals

Contents

5.1 Identification of Categories of Plastics
5.2 Thermoplastics
5.3 Cross-Linked Plastics (Elastomers, Thermosets)

Review Questions

Prerequisite Knowledge

What Are Plastics? (Lesson 1)
Raw Materials and Polymer Synthesis (Lesson 2)

5.1 Identification of Categories of Plastics

plastic categories

As we have seen, various bonding forces exist within a plastic. Plastics are classified according to macromolecular structure and types of bonding mechanisms. Figure 5-1 provides a summary and examples of these categories.

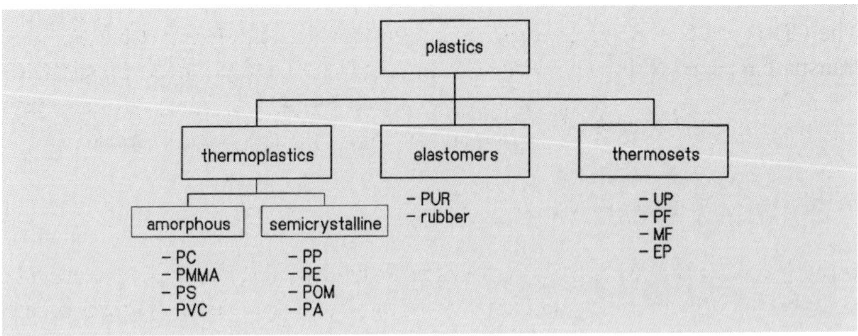

Figure 5-1 Classification of plastics

The four categories of amorphous thermoplastics, semicrystalline thermoplastics, thermosets, and elastomers are defined in greater detail.

5.2 Thermoplastics

definition

Plastics consisting of macromolecules with linear or branched chains held together by intermolecular forces are called "thermoplastics." The strength of the intermolecular forces depends on the type and number of branches or side chains, among other factors (Figure 5-2).

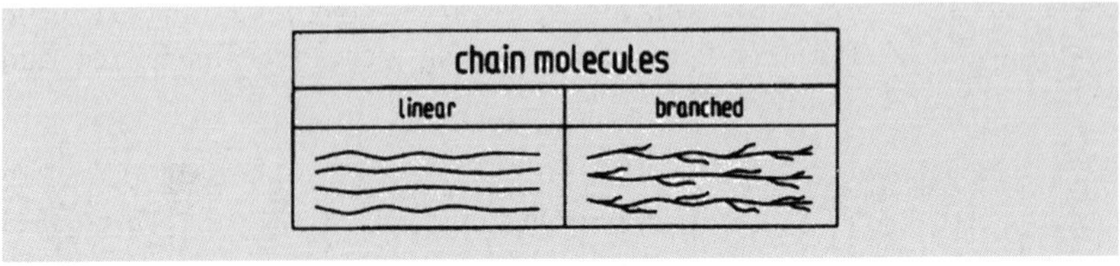

Figure 5-2 Linear and branched chain molecules

term

The term "thermoplastics" is derived from the words "thermos" (= hot, heat) and "plastic" (= shapable, moldable), because the application of heat weakens the intermolecular forces and makes the thermoplastic moldable.

Amorphous Thermoplastics

amorphous

Plastics with highly branched molecular chains and with long side chains cannot assume a tightly packed condition in any region because of their irregular structure. Chain molecules of this type are intertwined like a tangled ball or wad of cotton. The plastic is unstructured (= amorphous). It is therefore described as an "amorphous thermoplastic."

transparent as glass

Because amorphous thermoplastics are as transparent as glass (provided they have not been colored), these thermoplastics are also described as synthetic or organic glasses.

CD

The CD is made from an amorphous thermoplastic. Because this plastic is as transparent as glass, the laser can scan the depressions (pits) in the plastic with the help of the reflecting layer of aluminum or gold. The laser can then pass this information on to the CD player, which converts it back to music.

Semicrystalline Thermoplastics

characteristics

If the macromolecules exhibit only slight branching (i.e., only a few short side chains), or very ordered, regular branching (tacticity), some regions of the individual molecular chains will lie in an orderly arrangement, which allows them to be tightly packed. The regions in which the molecules exist in an orderly arrangement are described as regions of "crystallization," or "crystalline regions." The long molecular chains that become intertwined in the process of polymerization make it impossible to achieve complete crystallization, however.

crystalline

semicrystalline

Only certain portions of the chain molecules ever settle into a dense, orderly arrangement, whereas other portions are further apart and disarranged. These disorderly regions are called "amorphous regions." Thermoplastics containing both crystalline and amorphous regions are thus called "semicrystalline thermoplastics."

cloudy, milky

Even when they are not colored, semicrystalline thermoplastics are never as transparent as glass. Instead, the scattering of light that occurs at the boundaries between the amorphous and crystalline regions always makes the plastic somewhat cloudy or milky in appearance. Figure 5-3 is a schematic representation of the macromolecular arrangements in amorphous and semicrystalline thermoplastics.

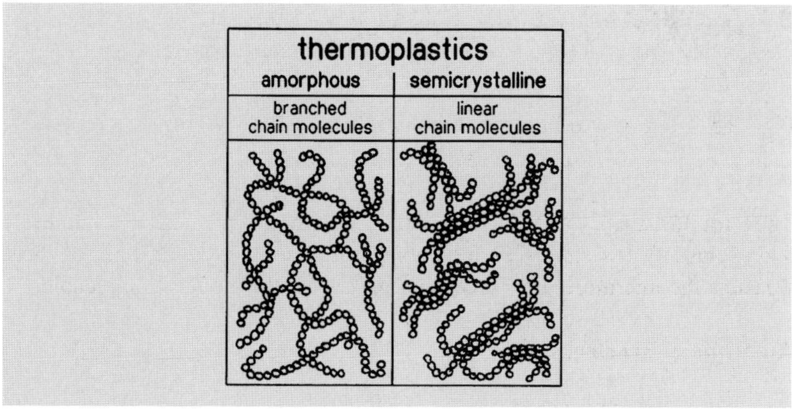

Figure 5-3 Structures of amorphous and semicrystalline thermoplastics

5.3 Cross-Linked Plastics (Elastomers, Thermosets)

cross-links

Aside from the category of thermoplastics, there are also categories of plastics in which the individual chain molecules are connected by transverse bonds (bridges). These transverse bonds (bridges) are also described as "cross-links," and the materials are accordingly called "cross-linked plastics." The categories, namely the elastomers and thermosets, are distinguished from each other by the respective abundance of cross-links in each. The molecules of these materials are therefore held together not only by intermolecular forces but also by atomic bonds.

Elastomers

characteristics

The chain molecules of elastomers are arranged randomly and exhibit relatively few cross-links. This category of polymers may be considered as lightly cross-linked, i.e., with a wide mesh.

properties

Elastomers behave like rubber at room temperature. The cross-links severely limit the ability of the individual molecular chains to move in relation to one another. Like the atomic bonds in the macromolecules, the atomic bonds in the bridges can only be broken by very high temperatures, and they are not reestablished when the temperature drops. Elastomers, therefore, do not melt nor are they soluble. Elastomers can be swelled to a certain extent, however, because the molecular chains exhibit few cross-links and therefore small molecules, such as oil or gasoline, can penetrate the spaces between the chains of the elastomer.

Thermosets Elastomers

characteristics

Thermoset elastomers combine easy processing of thermosets and characteristics of elastomers and are used more and more for new applications or substitute conventional thermoplast and elastomer material.

properties

Thermoset elastomer can repeatedly be melted and formed specifically during the cooling period. At room temperature, they react like rubber elastic material. Their structure and reaction lies in-between thermoplasts and elastomers. They are used widely in the car manufacturing industry.

Thermosets

characteristics

Another important category is represented by the thermosets, which also exhibit an irregular arrangement of the chain molecules. In comparison to the structure of the elastomers, however, they display a considerably greater abundance of cross-links between the individual molecule chains. Plastics constructed of such densely cross-linked chain molecules are called "thermosets."

properties

At room temperature, these densely cross-linked molecules are very hard and strong but brittle (i.e., sensitive to impacts). In comparison to thermoplastics, they display considerably less softening when heated. Like the elastomers, they do not melt and, because of the dense cross-linking, it is impossible to swell them.

Figure 5-4 is a schematic representation of the arrangements of macromolecules and their cross-links in elastomers and thermosets.

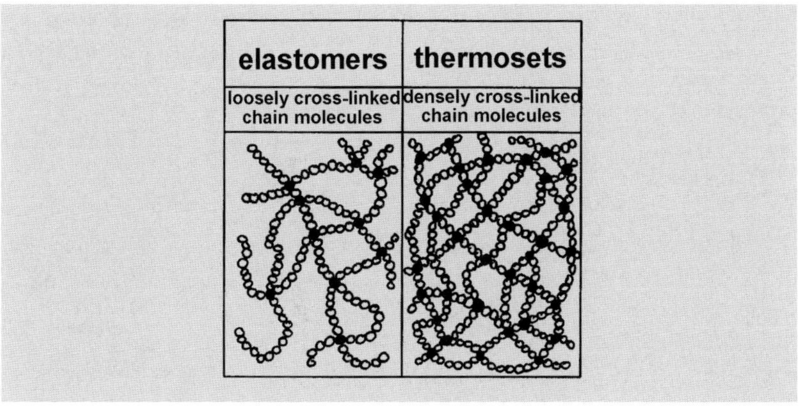

Figure 5-4 Structures of elastomers and thermosets

Review Questions

No.	*Question*	*Choices*
1	Thermoplastics are divided into amorphous and ____________________ thermoplastics.	thermosetting semicrystalline
2	Amorphous thermoplastics are ____________________ at room temperature.	cloudy glassy
3	The molecules of thermosets are ______________ cross-linked.	densely lightly
4	Elastomers have lightly cross-linked chain molecules and therefore ______________________ in a solvent.	dissolve swell
5	The CD consists of an amorphous thermoplastic, because the plastic must be _________________.	transparent swellable cross-linked
6	Thermoplastic elastomers can __________________ be melted.	not repeatedly

Lesson 6

Deformation Behavior of Plastics

Key Questions

How do plastics behave when heated?
How do amorphous and semicrystalline thermoplastics differ in this regard?
How do cross-linked plastics—elastomers and thermosets—behave?

Subject Area

Physics of Plastics

Contents

Prerequisite Knowledge

Bonding Forces in Polymers (Lesson 4)
Classification of Plastics (Lesson 5)

6.1 Behavior of Thermoplastics

deformation behavior

Deformation behavior is the manner in which the form of a molded part changes in response to a load (force) and temperature. Deformation behavior helps us explain the difference between a semicrystalline thermoplastic and an amorphous thermoplastic. We will illustrate this behavior by means of diagrams structured as in Figure 6-1. The diagram which corresponds with Figure 6-1 displays tensile strength and elongation at maximum tension as functions of temperature. The temperature scale is divided into the most important ranges for the respective plastic.

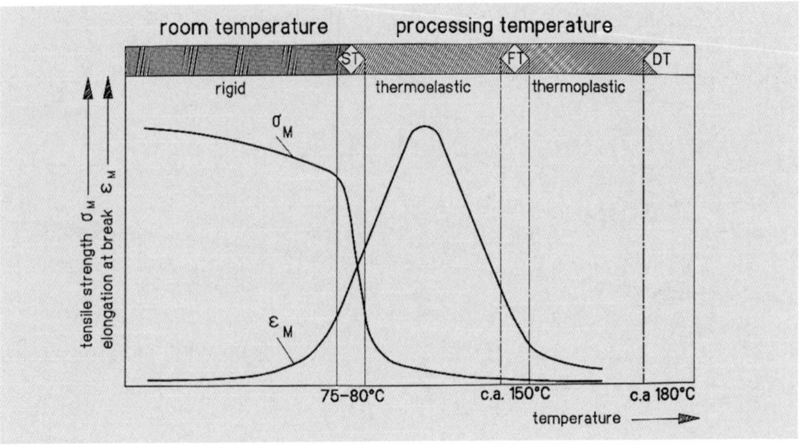

Figure 6-1 Diagram of deformation behavior

We will now explain tensile strength and elongation at maximum tension in greater detail. If a pulling force is applied to a plastic sample and then steadily increased, two effects are observed:

tensile strength

1. The plastic sample is able to withstand a certain maximum amount of stretching force. The amount of stress that exists at the maximum amount of force is called the "tensile strength." It represents a measure of the plastic's strength.

elongation at break

2. We also notice that the plastic sample becomes longer during the tensile test. In other words, it is elongated. The degree of elongation at which the plastic sample breaks is called "elongation at break." The ductility (toughness) of the plastic can be inferred from this measure.

effects of temperature

As mentioned, these two measured values depend on the temperature at which they are determined. We will now examine the deformation behavior of the different categories of thermoplastics.

6.2 Amorphous Thermoplastics

effects of temperature

We first consider the deformation behavior of an amorphous thermoplastic, as already shown in Figure 6-1. The plastic is a rigid material at room temperature. Because the individual macromolecules hardly move, they are held together by intermolecular forces. An increase in temperature causes the macromolecules to move with greater intensity. The strength of the material decreases, whereas its extensibility and ductility increase.

glass transition temperature

Once the temperature is greater than the softening temperature (ST), the intermolecular forces have become so weak that the influence of external forces can cause the macromolecules to slip apart from one another. The strength declines steeply and the elongation leaps upward. In this temperature range, the plastic exists in a rubber-elastic or thermoelastic state, in which it can be reshaped.

flow temperature range

As the temperature continues to increase, the intermolecular forces are eventually eliminated. The plastic proceeds in a continuous manner from the thermo-elastic state to the thermoplastic area. This transition is described as the range of the "flow temperature" (FT). This temperature cannot be specified precisely.

In the thermoplastic area, plastic pipes are manufactured by the extrusion process.

decomposition temperature

If the plastic continues to be heated, its chemical structure will eventually be destroyed. This limit is identified as the "decomposition temperature" (DT).

Rigid PVC (PVC-U) is an example of an amorphous thermoplastic. In Figure 6-2, the ranges of the different states are indicated as a function of temperature. Rigid PVC is used from approximately −10°C (14°F) to 50°C (122°F). This material changes to a thermoplastic state above a temperature of approximately 150°C (302°F).

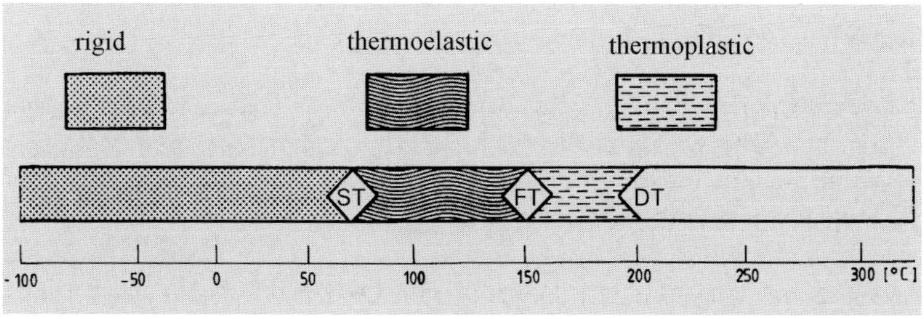

Figure 6-2 Temperature ranges for rigid PVC

6.3 Semicrystalline Thermoplastics

amorphous and crystalline area

As described above, the existence of two different regions side by side, so to speak, within a semicrystalline plastic distinguishes it from an amorphous plastic. One region is a crystalline region, in which the molecules are tightly packed; the other is an amorphous region, in which the molecules lie at a greater distance from one another. The intermolecular forces holding the crystalline region together are considerably stronger than those in the amorphous region. The temperature range within which the amorphous region of the plastic becomes thermoplastic is described as the "flow temperature" range (FT), and the corresponding range for the semicrystalline region is described as the range of the "crystalline melting temperature" (CMT).

The deformation behavior of a semicrystalline thermoplastic can be seen in Figure 6-3.

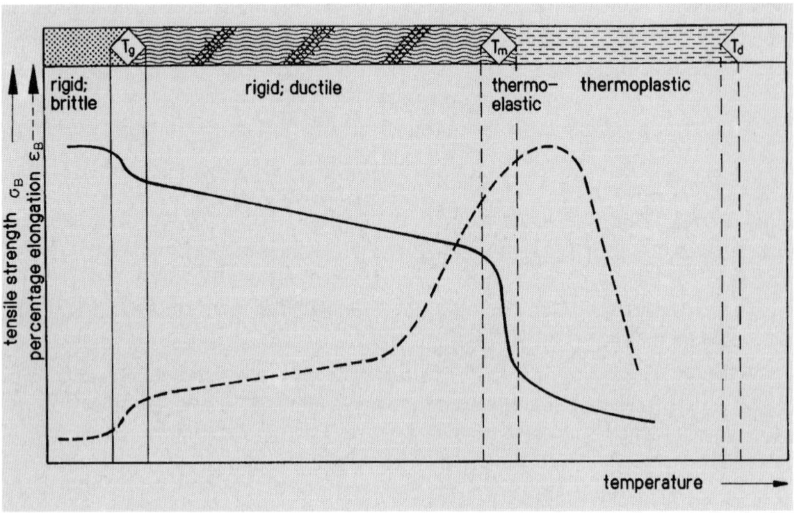

Figure 6-3 Deformation behavior of a semicrystalline thermoplastic

glass transition temperature

All regions of the plastic are stiff below the softening temperature (ST). As a result, the plastic is rigid and very brittle. Within this temperature range, the plastic cannot be used in practical applications.

temperature increase

As the temperature rises over the softening temperature (ST), the first effect is an increase in the mobility of the chain molecules in the amorphous regions. The reason for this is that the intermolecular forces are weaker in the amorphous regions than in the crystalline regions, where such forces remain strong. In conventional semicrystalline plastics, this temperature lies below room temperature. The plastic now exhibits both ductility and strength.

As the temperature rises, the mobility of the chain molecules in the amorphous regions continues to increase. The molecules in the crystalline regions also slowly begin to move. Soon, the crystalline melting temperature (CMT) is reached. At this temperature, the intermolecular forces are completely eliminated.

flow temperature range

Within the crystalline melting temperature (CMT), the semicrystalline thermoplant is thermoelastic and can be reshaped. Contrary to the amorphous thermoplast, the thermoelastic temperature area is very small and has to be kept rather exact during the reshaping process. As the temperature continues to increase, it approaches the crystalline melting temperature (CMT). At temperatures above the crystalline melting temperature, the bonding forces are too weak to prevent displacement and slippage of the chain molecules, even in the crystalline regions of the semicrystalline thermoplastic. The entire plastic now begins to melt. If heating is continued to a point above the decomposition temperature (DT), the plastic will be destroyed.

crystalline melting temperature

decomposition temperature

One example of a semicrystalline thermoplastic is high-density polyethylene (PE-HD or LDPE). The ranges of the most important states are shown in Figure 6-4 as a function of temperature. PE-HD (LDPE) is suitable for practical application within an approximate temperature range of approx. −15°C (5°F) and approx. 85°C (185°F).

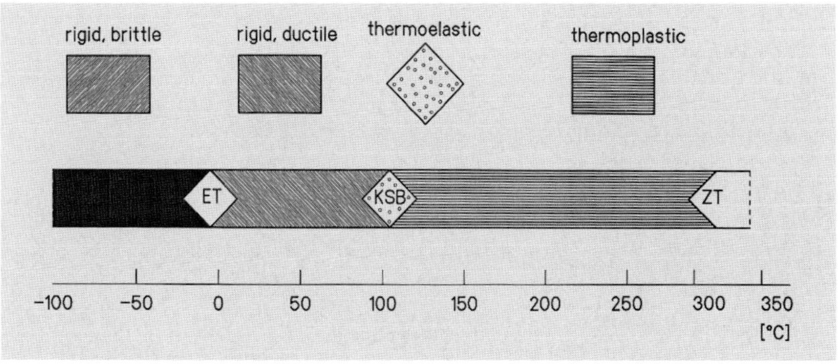

Figure 6-4 Temperature ranges for PE-LD (LDPE)

The amorphous thermoplastic PC, from which the CD is manufactured, can be used at temperatures up to 135°C (275°F). So even if the CD is left on a car dashboard and heated to 80°C (176°F) by direct sunlight, it will continue to function perfectly.

CD

6.4 Behavior of Cross-Linked Plastics (Elastomers and Thermosets)

shear modulus

rigidity

The deformation behavior of elastomers and thermosets can be best explained in terms of the torsion vibration test. In this test, the plastic's shear modulus G is measured. The shear modulus is a measure of the rigidity of the plastic. It is shown for plastics with varying degrees of cross-linking in Figure 6-5 as a function of temperature.

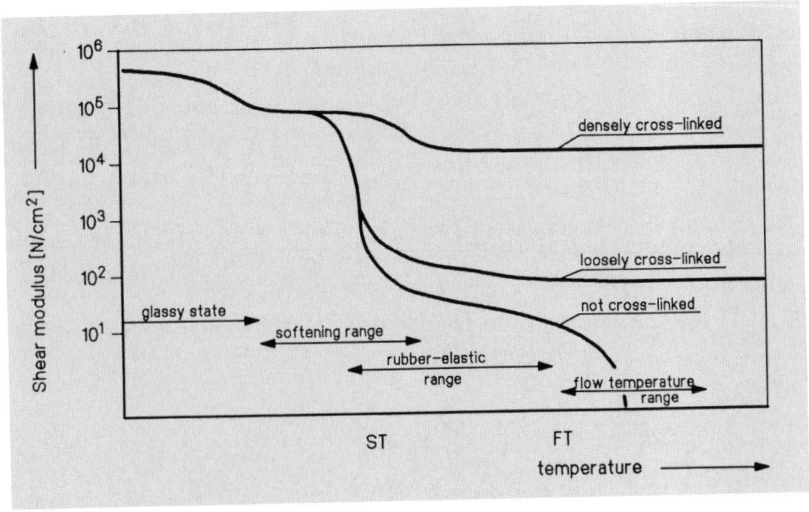

Figure 6-5 Shear modulus curves for cross-linked plastics

softening temperature

The plastic is rigid and brittle in the temperature range below the softening temperature (ST), regardless of how few cross-links there are.

elastomer

The shear modulus curve for a loosely cross-linked plastic (i.e., an elastomer) declines at temperatures immediately above the softening temperature. This means that the plastic retains only minimal rigidity at this point.

decomposition temperature

In contrast to thermoplastics, however, the loosely cross-linked plastic retains this rigidity through the flow temperature as the temperature continues to increase. This characteristic is attributable to the cross-links in the elastomer, which make it impossible for the individual molecular chains to slip away from one another. Therefore, the plastic will not melt, but will decompose if the temperature continues to rise beyond the decomposition temperature (DT).

temperature ranges

One example of an elastomer is natural rubber. Its temperature ranges are represented in Figure 6-6. Natural rubber can thus be used within a temperature range of –40°C to 130°C (–40°F to 266°F).

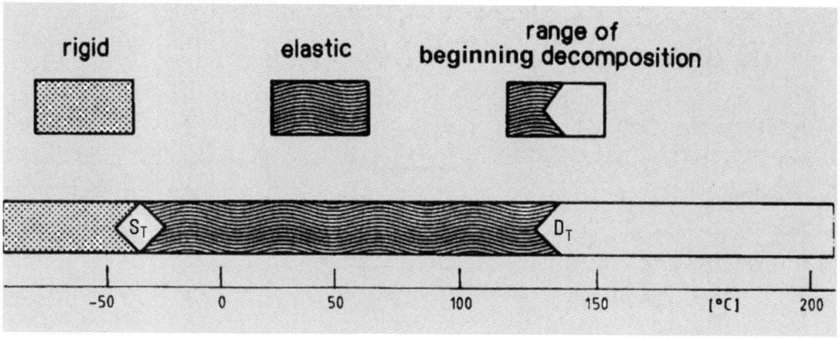

Figure 6-6 Temperature ranges for natural rubber

thermoset

The rigidity of densely cross-linked plastics (i.e., thermosets) decreases only slightly, even in the softening range. The abundance of cross-links between the individual chain molecules severely limits the ability of the macromolecules to move in relation to one another. Like elastomers, thermosets are infusible. They are also destroyed when heated beyond the decomposition temperature.

temperature ranges

The temperature ranges of a heat-resistant UP thermoset are represented in Figure 6-7 as an example. This thermoset can thus be used at temperatures below 170°C (338°F).

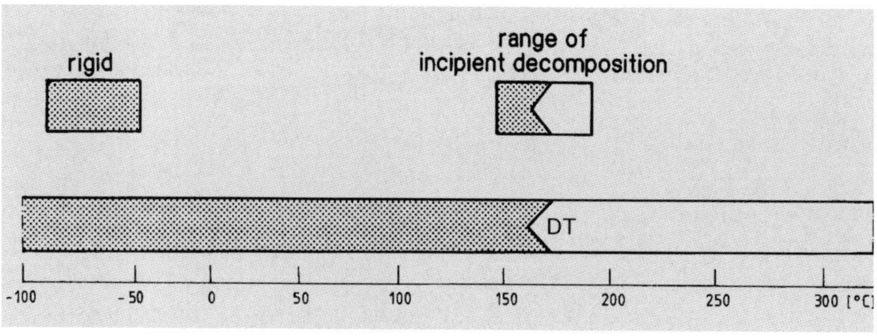

Figure 6-7 Temperature ranges for a thermoset made of UP

Review Questions

No.	*Question*	*Choices*
1	The elongation at the maximum tension is called the ____________________ ____________________.	tensile strength elongation at break
2	Tensile strength is a measure of the ____________________ of the plastic.	elasticity strength ductility
3	The elongation at break is an indication of the ____________________ of the plastic.	ductility tensile strength
4	Rigid PVC can be used within a temperature range of −10°C (14°F) to approximately ____°C, ____°F.	+50°C (122°F) +100°C (212°F) +150°C (300°F)
5	The softening temperature (ST) of semicrystalline thermoplastics usually lies ____________________ room temperature.	below above
6	The semicrystalline thermoplastic PE-HD (LDPE) can be used within an approximate temperature range of ____°C (____°F) to 85°C (185°F).	−15°C (5°F) +15°C (60°F)
7	CDs are made of the ____________________ thermoplastic polycarbonate (PC) because they must be transparent.	amorphous semicrystalline
8	The shear modulus of a cross-linked plastic is a measure of its ____________________.	rigidity strength ductility
9	Elastomers and thermosets are infusible because they are ________.	cross-linked amorphous
10	Natural rubber can be used within an approximate temperature range of −40°C (−40°F) to ______°C, ______°F.	+40°C (105°F) +130°C (265°F) +180°C (355°F)

Lesson 7

Time-Dependent Behavior of Plastics

Key Questions

How does the strength of a loaded plastic change over time?
What is the "creep" of a plastic?
How do dependence on time and temperature affect the strength of plastics and, in turn, the use of these materials?

Subject Area

Physics of Plastics

Contents

Prerequisite Knowledge

Bonding Forces in Polymers (Lesson 4)
Deformation Behavior of Plastics (Lesson 6)

7.1 Load-Bearing Behavior of Plastics

tensile test

In a tensile test, we simultaneously apply an identical load to a plastic sample and to a metal sample. The samples become elongated as shown in Figure 7-1. If the load were immediately removed from the samples, each sample would return to its original length.

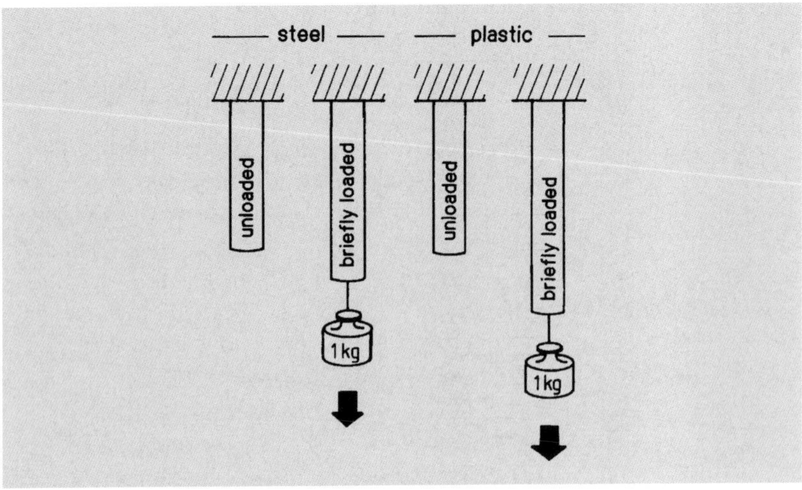

Figure 7-1 Deformation of samples after brief loading (1kg = 2.2lb)

stress
strain

modulus of elasticity

Let us now maintain the load on the samples and measure their elongation, or "strain." The force of the load upon the sample is divided by the cross-sectional area of the sample. This yields the amount of stress acting upon the sample. If we then divide the stress by the strain, we obtain the modulus of elasticity for the sample. This measure thus indicates the amount of strain that a material will undergo when subjected to a certain load; in other words, it is a measure of the material's strength. The higher the modulus of elasticity, the less strain is exhibited by a material at a given load, and the greater is its rigidity. Figure 7-2 shows the modulus of elasticity for various materials.

	modulus of elasticity	
material	**[N/mm^2]**	**psi**
plastics	200–15,000	30,000–2,000,000
steel	210,000	30,000,000
aluminum	50,000	7,250,000

Figure 7-2 Modulus of elasticity for various materials

It can be seen that the modulus of elasticity (and therefore the rigidity) of steel is up to 1000 times greater than that of plastic. For this reason, the length of the steel sample changes less than the length of the plastic sample when both are subjected to identical loads for brief periods (Figure 7-1). When technical components are constructed from metal, the modulus of elasticity obtained from such short-term testing is critically important to the design. This measure plays only a secondary role in the design of plastic parts, however. It allows us to make only conditional conclusions about the strength of the plastic, since the rigidity of plastics depends on time.

comparison: plastic vs. steel

7.2 Effects of Time on Mechanical Behavior

Let's take another look at the two samples discussed. By now, both of them have been subjected to a constant load for a somewhat longer period of time. If we measure the strain of the samples once again, we find that the metal sample displays the same amount of strain as before (Figure 7-3). But the strain of the plastic sample has increased, although the load has remained the same. This is a typical characteristic of plastics known as "creep".

creep

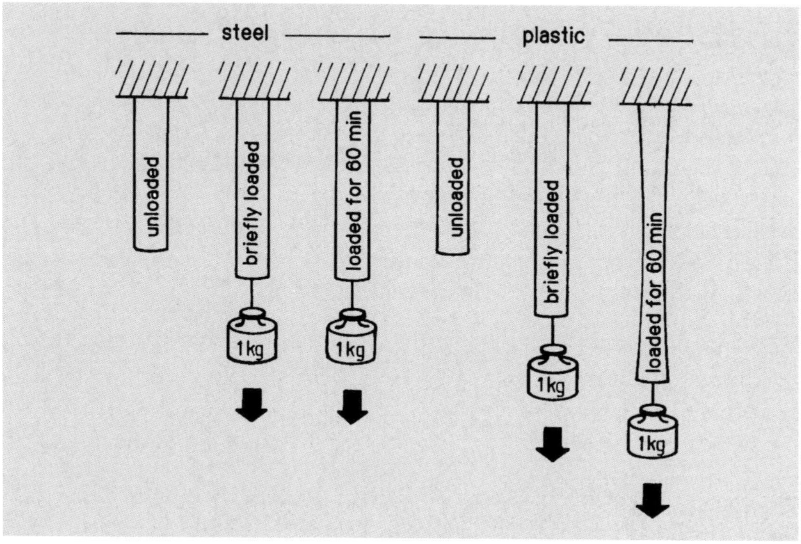

Figure 7-3 Deformation of samples after extended loading

The creep of a plastic can be explained in terms of its internal structure. As we have shown, plastics consist of tangled macromolecules held together by intermolecular forces. When a plastic is subjected to a load, this tangled mass at first becomes elongated.

internal structure

strain

If the load is relatively small and very short in duration, this strain disappears. If the plastic is loaded for a longer period, the intermolecular forces slowly begin to weaken. The macromolecules slip apart from one another. The plastic does not recover from the resulting strain, even when the load is removed.

slippage of macromolecules

viscoelasticity

Maxwell model

The strain of the plastic is partially elastic (the elongation of the tangled mass) and partially viscous and permanent (the slippage of the molecules). The plastic is therefore described as behaving in a "viscoelastic" manner. This behavior is represented by the Maxwell model (see Figure 7-4). The model consists of a dashpot and a spring. When the model is loaded with a force, the spring spontaneously extends, but the dashpot does not yet react. If the load is maintained, the dashpot slowly begins to elongate. When the load is removed from the model, the spring spontaneously recovers from its strain, but the strain of the dashpot remains as permanent strain.

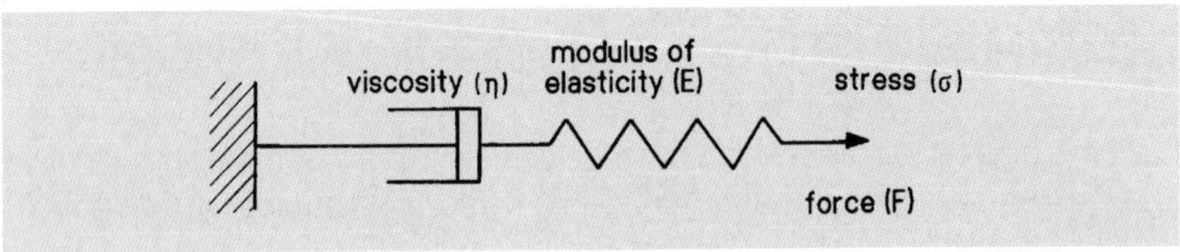

Figure 7-4 Maxwell model

7.3 Recovery Behavior of Thermoplastics

recovery effect

As described, a tangle of macromolecules becomes somewhat elongated when it is subjected to a load. In other words, the macromolecules are stretched to a greater length. If the load is immediately removed from the plastic, the molecules return to their original position; that is, the strain is eliminated. A behavior of plastics that is based on the same principle is called the "recovery effect." We consider a simple plastic tube in which the macromolecules exist in a tangled state (Figure 7-5, position a).

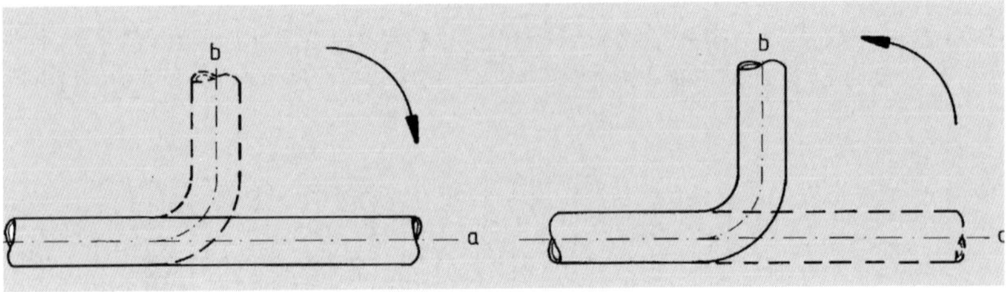

Figure 7-5 Recovery behavior of a plastic tube

thermoelastic range

forming

Let us heat the tube until its temperature reaches the thermoelastic range (above the softening temperature ST but below the flow temperature range FT). It is now relatively easy to bend the tube into a right angle (Figure 7-5, position b). In the jargon of plastics processing, this process of heating and deformation is also called "forming." After forming is complete, we quickly cool the tube to a point below the softening temperature. The tube remains deformed.

If we could examine the chain molecules at the bend in the tube, we would find that they are no longer entangled in their original state. Instead, the molecules at this location exist in a stretched form. This condition is described as "orientation" within the plastic. The temperature is too low to allow the molecules to move back to their entangled original form. We say that the orientations are frozen in.

orientation

If we now reheat the deformed tube, the chain molecules move back to their starting positions, thus pulling the tube back into its original straight form (from position b to position a in Figure 7-5). The orientation disappears. This process is called "recovery behavior."

recovery behavior

7.4 Dependence of Plastics on Temperature and Time

As explained, temperature and time have a critical influence on the mechanical behavior of plastics. In the design of plastic parts, a decisive role is therefore played by the temperature and duration of loading that will later be encountered in use. These factors are less important in the design of metal parts.

design

In order to aid the designer in estimating these two influences, "creep curves" are established for individual plastics (see Figure 7-6). A plastic sample with a predefined cross-sectional area is loaded with a force at a certain temperature. The change in strain is then measured over time. The test is repeated at different levels of force and different temperatures. The recorded measurement values are plotted as curves on a diagram. The diagram (Figure 7-6) shows the strain of the samples as a function of the duration of loading. Each of the curves represents a certain load, or stress (σ), and a certain temperature. The symbol σ_1 signifies the lowest applied stress, and σ_4 signifies the highest applied stress.

creep curves

measurements

diagrams

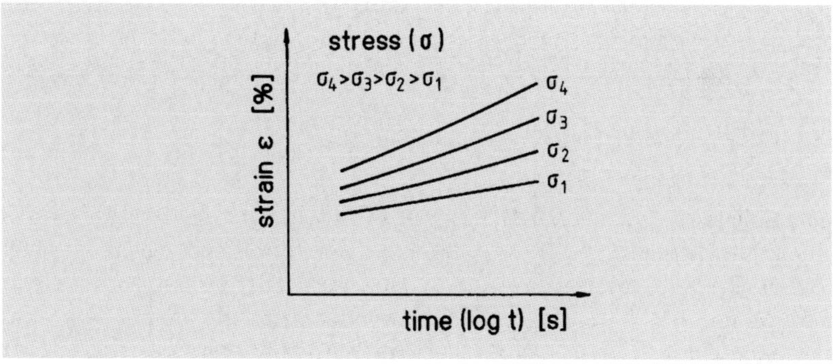

Figure 7-6 Creep curves

In order to simplify these diagrams, they often show only those curves that apply to a certain temperature, for example. On a diagram of this type, it is thus possible to clearly discern the changes in the curves that can be attributed to differences in load. The loads are specified as stresses—that is, as the load applied to a given unit of cross-sectional area. This makes it possible to use the values in designing parts of any desired cross-section. However, the information contained in a creep curve diagram would often be more useful to the designer if it were presented in a different way. In such cases, the diagrams described are converted to other forms.

designing

One of these forms is the time-dependent creep diagram. On a diagram of this type, the stress is shown at a constant temperature and strain (ϵ) as a function of time (see Figure 7-7). This kind of creep diagram is especially useful for obtaining the allowable stresses, and thus the allowable loads, in situations where it is not permissible to exceed a certain amount of strain (elongation) for the part being designed.

time-dependent creep diagram

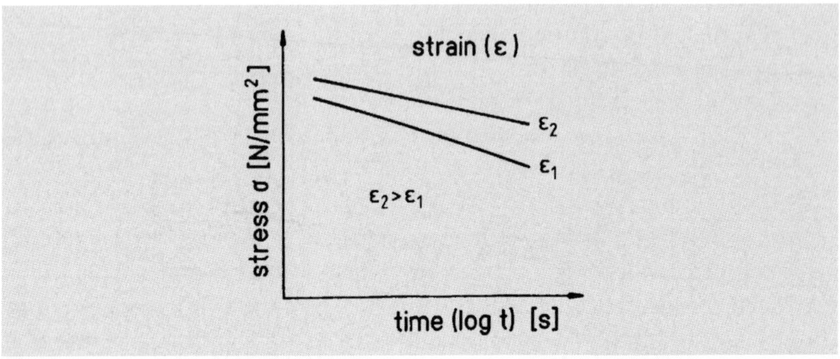

Figure 7-7 Time-dependent creep diagram [$1\ \text{N/mm}^2 = 1.45 \times 10^2$ psi(lb/in^2)]

Another type of diagram is the isochronous stress-strain diagram. In this case, the stress is shown at a constant load time (iso = equal, chronos = time) and at a constant temperature as a function of strain (see Figure 7-8). This diagram is especially useful for finding the range within which strain is linearly dependent on stress for the respective plastic.

stress-strain diagram

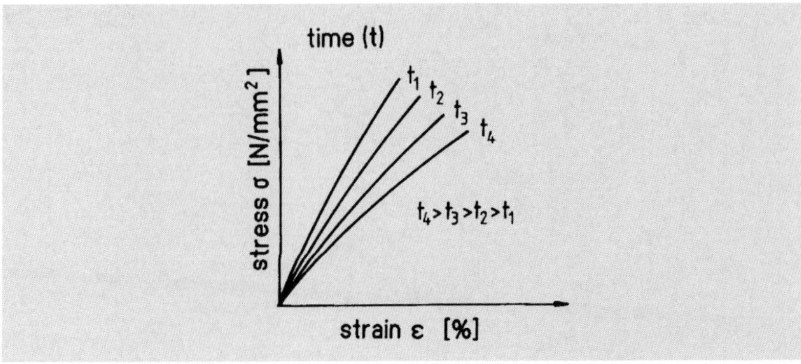

Figure 7-8 Isochronous stress-strain diagram [$1\ \text{N/mm}^2 = 1.45 \times 10^2$ psi]

Now let's take a look at how the creep diagram is used. Figure 7-9 shows a creep diagram for PMMA at 23°C (73°F). We assume that a component will be subjected to a stress of 40 N/mm^2 (5800 psi) at a temperature of 23°C (73°F). How long would it take for the component to break? How long would it take if the stress were changed to 50 N/mm^2 (7250 psi) if the other assumptions remained unchanged?

example

For a stress of 40 N/mm^2 (5800 psi), we obtain a duration of approximately 10^4 hours. This works out to a period of approximately 14 months. For a stress of 50 N/mm^2 (7250 psi), we obtain a duration of 2.6×10^2 hours, or approximately 11 days! This clearly demonstrates the extremely time-dependent nature of the behavior of plastics. It also demonstrates that this consideration must not be neglected in the design process.

result

time dependence

design process

Now let's examine PMMA at the same load (40 N/mm^2) (5800 psi) but at varying temperatures. As we have seen, PMMA will break after about 14 months at a temperature of 23°C (73°F). If we increase the temperature to 60°C (140°F), we find that it no longer takes 14 months, but only 1.5 hours (90 minutes) for the plastic to break. The dependence on temperature is therefore even more drastic than the dependence on time.

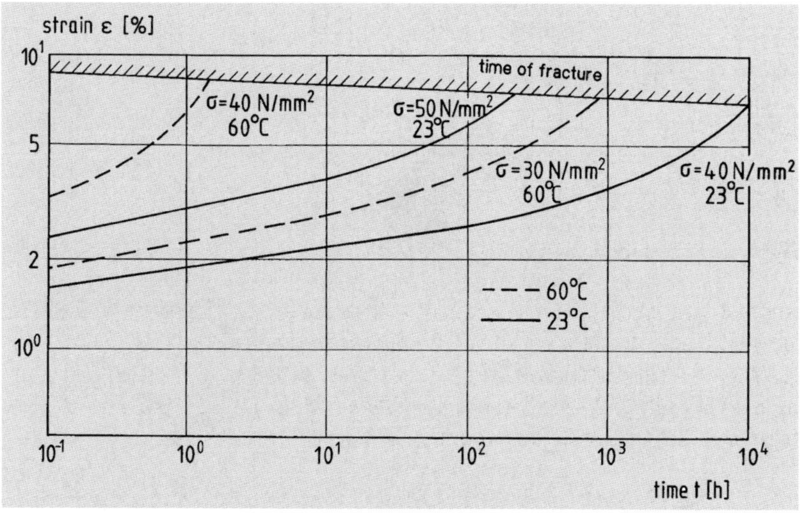

Figure 7-9 Temperature dependent creep diagram for PMMA

Review Questions

No.	*Question*	*Choices*
1	The modulus of elasticity is a measure of the ________________ of a material.	strength rigidity plasticity
2	The modulus of elasticity is up to ____________________ times higher for steel than for plastics.	10 100 1000
3	The strain of a plastic ____________________ dependent on the magnitude and duration of the load.	is not is
4	The change in the strain of a plastic under a constant load is called ____________________.	slippage creep
5	The recovery effect can become noticeable when formed plastics are ____________________.	heated cooled
6	The existence of stretched, frozen-in-position chain molecules is called ____________________.	orientation stress
7	The dependence of the mechanical behavior of plastics on ____________________ must be considered in the design process.	time and temperature temperature time
8	The designer of plastics parts is aided in his or her work by diagrams, such as creep curves, ____________________ and isochronous stress-strain diagrams.	time-dependent creep diagrams endurance diagrams
9	The strain (ϵ) of PMMA has a value of approximately _______% after 14 days at a load of 30 N/mm^2 (4350 psi) and a temperature of 60°C (140°F). (Determine value from Figure 7-9).	2 5 10
10	After a period of approximately __________ [h] at a load of 50 N/mm^2 (7250 psi) and a temperature of 23°C (73°F), the strain (ϵ) of PMMA attains a value of 5%. (Determine value from Figure 7-9).	10 50 100

Lesson 8

Physical Properties

Key Questions

How does the density of plastics compare to the density of metals?
How well do plastics conduct heat?
How well do plastics conduct electricity?
What are the optical properties of plastics?

Subject Area

Physics of Plastics

Contents

Prerequisite Knowledge

Plastics Fundamentals (Lesson 1)

8.1 Density

density range

The very low density of plastics distinguishes them from other materials (see Figure 8-1). The density range of plastics extends from approximately 0.9 to 2.3 g/cm^3. Examples of low-density plastics include the bulk plastics polyethylene (PE) and polypropylene (PP). Both of these materials have a lower density than water and therefore float. As a result of their greater buoyancy, it is possible to separate these two plastics from heavier plastics in water. Most plastics lie within a density range of 1 to 2 g/cm^3. Only a few plastics (for example, polytetrafluoroethylene (PTFE)) have a density greater than 2 g/cm^3.

material	**density** ρ**[g/cm³]**
plastics general	0.9–2.3
PE	0.9–1.0
PP	0.9–1.0
PC	1.0–1.2
PA	1.0–1.2
PVC	1.2–1.4
PTFE	>1.8
steel	7.8
aluminum	2.7
wood	0.2–0.95
water	1.0

Figure 8-1 Density of various materials

explanation

The density of metals can be several times greater than the density of plastics. For example, the density of aluminum is approximately 2.7 g/cm^3, and the density of steel is 7.8 g/cm^3. The higher density of metals can be attributed to two factors:

1. Their individual atoms (aluminum, iron) are heavier than the atoms from which plastics are constructed, namely, carbon, nitrogen, oxygen, and hydrogen.

2. The average distance between the atoms is greater in plastics than in metals.

8.2 Thermal Conductivity

thermal conductivity

Thermal conductivity is a measure of how well a material can convey heat. The thermal conductivity of plastics lies within a range of 0.15—0.5 W/mK. This is a very low value. The thermal conductivity values of other materials are compared to plastics in the list in Figure 8-2. For example, metals can display values up to 2000 times higher than the values for plastics. Metals conduct heat very well. On the other hand, air conducts heat 10 times less effectively than plastics.

material	thermal conductivity λ [W/mK]
plastics general	0.15–0.5
PE	0.32–0.4
PA	0.23–0.29
steel	17–50
aluminium	211
copper	370–390
air	0.05

Figure 8-2 Thermal conductivity of various materials

explanation

One reason for the low thermal conductivity of plastic is the lack of freely mobile electrons in the material. Because metals do possess such mobile electrons, they are good conductors of both electrical current and heat.

processing

One disadvantage of poor thermal conductivity becomes apparent in the processing of plastics. The heat required for processing can only be slowly introduced into the plastic, and it is difficult to dissipate this heat upon completion of processing.

applications

However, the tendency that proves disadvantageous in processing often turns out to be an advantage in daily use. For example, pot handles are made of plastics. They do not become hot as quickly as metals when the pot is heated. This allows us to remove the pot from the stove without burning our fingers. Plastics are also used as insulating materials in the construction industry. Air is “admixed” with the plastic, because air conducts heat even less effectively than plastic, as mentioned. This produces a foamed plastic with a thermal conductivity that represents an average of the values for air and plastic. On the other hand, it is possible to add metallic fillers to the plastic in order to increase its thermal conductivity. The effects of various mixtures on thermal conductivity can be determined from Figure 8-3.

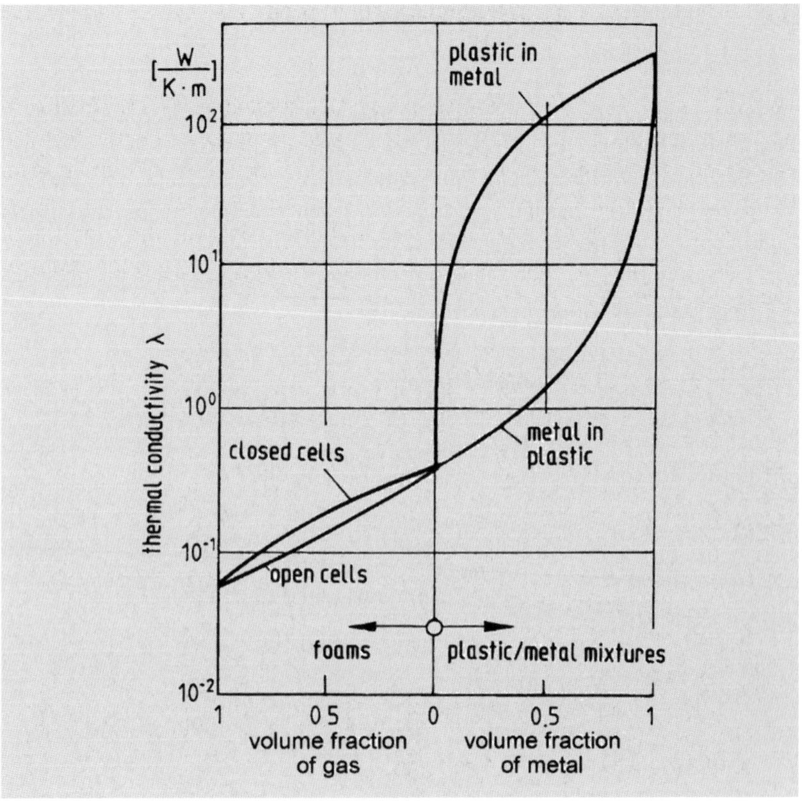

Figure 8-3 Thermal conductivity of plastic mixtures

8.3 Electrical Conductivity

electrical conductivity

Electrical conductivity is a measure of how well a material can carry electric current. Plastics generally do a very poor job of conducting electric current. They exhibit high resistance, and therefore low conductivity, in comparison to other materials (see Figure 8-4). The electrical resistance of plastics depends on temperature. As the temperature increases, the resistance decreases and the plastic becomes more conductive.

material	**electrical conductivity [m/ohm mm^2]**
PVC	10^{-15} [up to approx. 60°C (140°F)]
steel	5.6
aluminum	38.5
copper	58.5

Figure 8-4 Electrical conductivity

One reason for the low electric conductivity of plastics is their lack of the free electrons which occur in metals.

explanation

If it is desirable for a plastic to achieve greater conductivity, it is possible to incorporate metal powder into the plastic. The resulting effect on the electrical resistance of the plastic can be determined from Figure 8-5. As can be seen, the resistance decreases by a factor of 10,000,000 (ten million!) when 20 volume % of metal is mixed into the plastic. Because of their very high electrical resistance, plastics are preferred as insulating materials for electrical devices and wires.

enhancement of conductivity

insulation

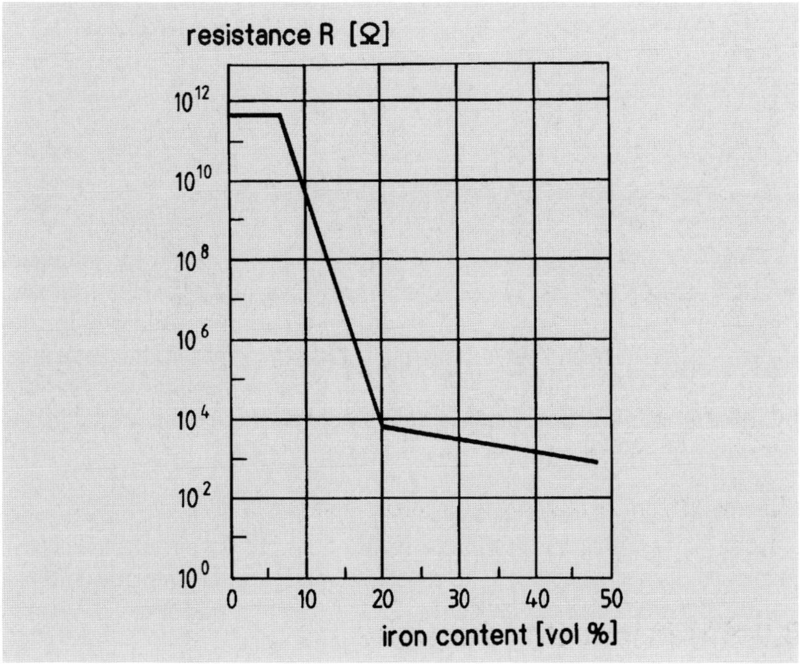

Figure 8-5 Resistance of a plastic with metal powder filler

8.4 Transparency

When light strikes a material, only a certain portion of this "incident" light passes through the material without being reflected or absorbed. The ratio of the intensity of this "transmitted" light to the intensity of the incident light is called "transparency" or "transmittance." Amorphous thermoplastics such as PC, PMMA, PVC, and UP resin are essentially as transparent as window glass. Their transparency measures approximately 90% (see Figure 8-6). This corresponds to a transmittance value of 0.9%. In other words, factor 0.1 or 10% of the light is lost through reflection and absorption.

transmittance

However, one disadvantage of plastics is that environmental influences, such as weathering and stresses produced by changes in temperature, can cause hazing, which in turn impairs transparency.

environmental influences

material	transparency [%]
PC	72–89
PMMA	92
window glass	90

Figure 8-6 Transparency

CD

CDs are manufactured from the amorphous thermoplastic polycarbonate (PC), because this plastic is characterized by a high degree of transparency, in addition to its other favorable properties.

8.5 Characteristic Material Values

property classes

The physical properties presented here do not represent all of the properties that a plastic can possess. The various properties can be divided into classes, such as mechanical properties and thermal properties. The various properties of each class are almost always accompanied by several physical measurement values, each of which describes the plastic in terms of the respective property. By using these measured values, a designer or planner can select the plastic that meets his or her requirements.

databases

In addition to sets of tables and handbooks, it also possible to use the computer to facilitate the search for the right plastic. Various plastics manufacturers have developed plastics databases. The measured values for the properties of most plastics are stored in these databases.

search profile

The databases are so convenient that the user merely needs to enter the required minimum and maximum values for certain properties—the "search profile." The computer locates all plastics that display a combination of these desired properties. The user can then view the property lists for the individual plastics that the computer has found. Figure 8-7 shows a database excerpt of the mechanical properties of acrylonitrile/butadiene/styrene (ABS) as they appear on a computer screen.

explanation

This is an ABS plastics material of Bayer AG which is called Novodur P2H-AT. One can see some mechanical properties.

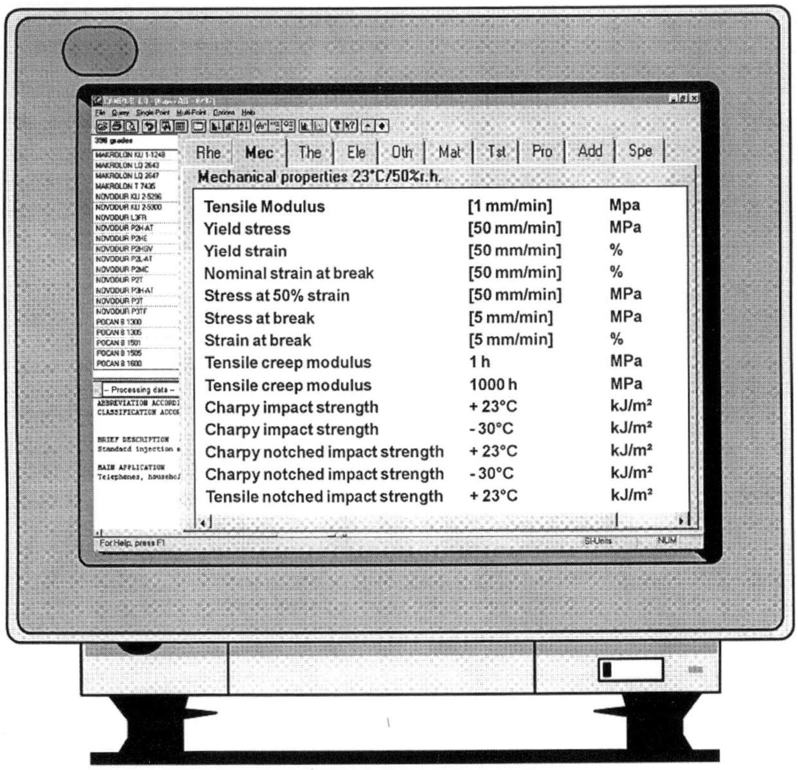

Figure 8-7 Screen page from a materials data base (CAMPUS)

Review Questions

No.	*Question*	*Choices*
1	Plastics are generally ______________________ than metals.	lighter heavier
2	The density of steel measures 7.8 g/cm^3. The densities of plastics lie within a range of ______________________ g/cm^3.	0.5 to 0.8 0.9 to 2.3 2.5 to 5.0
3	Metals are up to ______________________ times more thermally conductive than plastics.	20 200 2000
4	The poor electrical conductivity of plastics can be enhanced by additives such as ______________________.	chalk flour metal powder glass splinters
5	The transparency of amorphous thermoplastics is _______________ the transparency of glass.	greater than less than roughly equal to
6	CDs consist of PC because of the favorable ___________________ of this amorphous plastic.	thermal conductivity transparency density

Lesson 9

Plastics Fabrication and Processing

Key Questions What processing methods are used with plastics?
What fabrication methods are used with plastics?

Subject Area From Raw Plastic to Finished Product

Contents

Prerequisite Knowledge Classification of Plastics (Lesson 5)
Deformation Characteristics of Plastics (Lesson 6)

9.1 Methods of Fabrication and Processing

raw plastic

In order to get from the raw plastic material manufactured by chemical processes to the plastic product used by the consumer, some intermediate steps are required. The basic substance for the plastic is manufactured in the form of grains ("granulate" or resin), powder, paste, or liquid. It is then processed into semifinished products or finished articles.

semifinished products

finished products

Semifinished products are intermediate products that await further conversion to finished products by various fabrication techniques (e.g., reshaping). Examples of semifinished products include plastic sheets, films, tubes, and profiles. Finished articles, or finished products, are manufactured by primary forming processes (e.g., the method of "injection molding"). Examples of finished products include plastic buckets, gears, and housings.

overview

An overview of fabrication and processing methods for thermoplastics and thermosets is given in Figure 9-1.

processing and shaping techniques	**thermosets**	**thermoplastics**
primary processing	molding compounds are shaped at the same time that a chemical reaction occurs: - thermosetting molding compounds - liquid reaction resins	Molding compounds are shaped while in a thermoplastic state
thermoforming		semifinished products are shaped while in a thermoelastic state.
machining	shaping by cutting operations	shaping by cutting operations
joining	mechanical joining methods cementing (adhesive bonding)	mechanical joining methods cementing (adhesive bonding) welding

Figure 9-1 Overview of fabrication and processing methods

thermosets

You will notice that this overview does not specify a reshaping method for thermosets. (This also applies to elastomers.) Cross-linked plastics do not exist in a thermoplastic state in any temperature range. Therefore, they can no longer be reshaped after the curing process.

machining

"Machining" is a generic term for plastic shaping methods involving cutting operations, such as turning on a lathe, milling, and sawing.

"Joining" is a generic term for methods of attaching plastics. These methods include cementing and welding, as well as mechanical methods of attachment, such as screws and rivets.

joining

Reshaping, machining, and joining are combined under the general heading of "fabrication" methods, whereas primary forming processes (such as extrusion and injection molding) are considered "processing" methods.

fabrication
processing

9.2 Methods for Shaping Thermoplastics

In Figure 9-2, the various processing and shaping methods are listed under the respective material states of thermoplastics. This classification does not apply to cross-linked plastics (thermosets and elastomers). If an article manufactured from a thermosetting or an elastomeric material is not already in its final form when cross-linking is complete, it can only be fabricated by machining or by joining. It is not even possible to weld these materials because they do not assume a thermoplastic condition in any temperature range.

processing and shaping techniques	**material state**		
	rigid	**thermoplastic**	**thermoplastic**
primary processing			extrusion casting calendering injection molding compression molding
reshaping		folding/bending embossing/knurling stretch forming vacuum forming combined methods	
machining	drilling lathe turning milling planing sawing cutting grinding		
joining	screwing riveting cementing (adhesive bonding)		welding

Figure 9-2 Classification of processing and shaping methods

Review Questions

No.	*Question*	*Choices*
1	Injection molding belongs to the category of __________________.	reshaping primary processing joining machining
2	Thermoforming belongs to the category of ____________________.	reshaping primary processing joining machining
3	________________________ is a method of joining.	cementing extrusion thermoforming
4	________________________ is a method of machining.	milling welding cementing
5	Thermosets and elastomers cannot undergo ____________________ because they do not enter a thermoplastic state when heated.	cementing thermoforming milling
6	Which method of joining can be used with thermoplastics and with thermosets? ________________________	cementing welding mechanical joining

Lesson 10

Plastics Compounding

Key Questions

Why are plastics compounded?
What are the functions of the individual additives?
What steps are involved in compounding?

Subject Area

From Raw Plastic to Finished Product

Contents

Prerequisite Knowledge

Raw Materials and Polymer Synthesis (Lesson 2)
Physical Properties (Lesson 8)

10.1 Overview

compounding

We have seen how a plastic is made from a raw material. To ensure that this plastic will exhibit good processibility, as well as properties appropriate to its later use, we must compound it. Compounding provides the plastic with the necessary properties for processing and use. Figure 10-1 gives an overview of the various types of compounding.

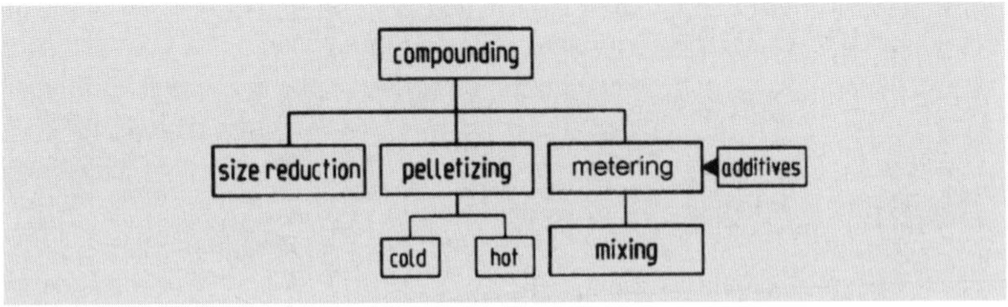

Figure 10-1 Types of compounding

functions

Compounding performs two important functions. For one thing, it ensures that the additives, which may be present in widely disparate quantities, are distributed evenly throughout the overall material. Second, compounding converts the raw plastic to a form that facilitates processing (e.g., a granulate).

10.2 Additives and Metering

Additives

properties

By purposefully incorporating additives into the plastic, we can alter its properties (see Figure 10-2).

additives	**effect**
antioxidants (heat stabilizers)	inhibit reactions that cause degradation of the plastic through oxidation
light stabilizers	inhibit reactions that cause degradation of the plastic through exposure to light (ultraviolet light)
lubricants	influence processing properties of the plastic during plasticizing
plasticizers	lower the modulus of elasticity
pigments	color the plastic
reinforcing agents	raise the modulus of elasticity

Figure 10-2 Additives used in plastics processing

We will now examine the effects of such additives in greater detail, using heat stabilizers and plasticizers as examples. A heat stabilizer allows the plastic to withstand the required processing temperature without being degraded. This additive thus facilitates the processing of the plastic.

heat stabilizers

Plasticizers lend flexibility and extensibility to plastics that are rigid and brittle by nature. This makes it possible for these plastics to penetrate entirely new areas of application. For example, it is possible to make a flexible, tough film from a plastic that would otherwise be rigid and brittle. This additive thus alters the performance characteristics of the plastic.

plasticizers

Metering

Because successful compounding of additives with the raw plastic depends on precise metering of the individual components, it is necessary to measure (meter) these components. Metering can be performed in two ways. The substances can be metered by volume ("volumetric feeding") or by weight ("weight feeding").

types of metering

The disadvantage of metering by volume is a relative lack of precision. The reason for this imprecision is that the substances usually exist in a granular state. The spaces between the grains vary in size. As a result, similar volumes actually represent different amounts of a given substance in many cases. The advantage of this method is the relatively low price of the metering equipment.

volumetric feeding

The method of metering by weight—in other words, weighing—is considerably more precise and much easier to automate than the method of metering by volume. Unfortunately, the required equipment is much more expensive.

weight feeding

10.3 Mixing

The goal of mixing is to distribute the additives as evenly as possible throughout the plastic, without subjecting the plastic to excessive stress. As a rule, mixing occurs in machines that function in a batchwise manner. These machines produce relative motion between the particles of the substances being mixed. A distinction is made between two methods of mixing, namely, cold and hot mixing.

mixing methods

Cold Mixing

Cold mixing is performed at room temperature. The individual components are merely blended together. One example of this method is the free-falling mixer (Figure 10-3), in which mixing is performed entirely by the effect of gravity. It is especially well suited for mixing substances of different grain sizes.

cold mixing

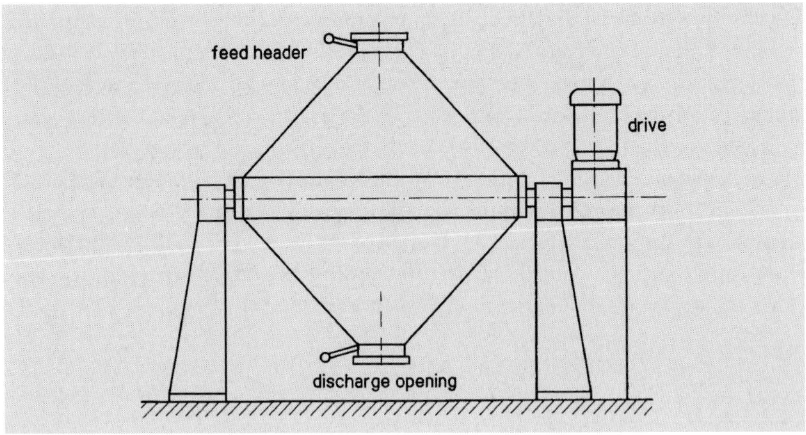

Figure 10-3 Free-falling mixer

Hot Mixing

hot mixing

turbulent mixer (Wirbel mixer)

Hot mixing involves the heating of the material being mixed. At temperatures up to 140°C (285°F), certain additives melt and diffuse into the plastic. An example of a hot mixer is the "Wirbel mixer or turbulent mixer", consisting of a heating mixer and a cooling mixer (see Figure 10-4). The high-speed agitator of this mixer produces a powerful relative motion between the particles of the material being mixed. The material is fused by the resulting heat of friction and, in some cases, by external heating. The completely mixed material is transferred from the heating mixer to the cooling mixer for purposes of storage.

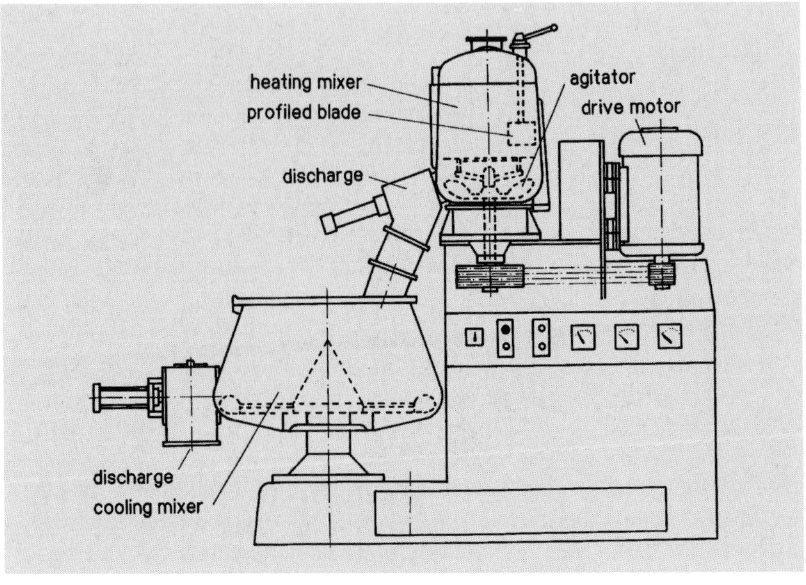

Figure 10-4 Turbulent mixer (Wirbel mixer)

10.4 Plasticizing

homogenization

fillers

Plasticizing converts the completely mixed plastic to a form amenable to further processing. Another effect of plasticizing is an increased homogenization of the plastic. Large quantities of additives (fillers) can also be added in this step of the operation. This would be uneconomical in the hot mixer. Roll mills, kneaders, and screw machines are suitable for this operation.

shearing roll mill

Figure 10-5 shows an example of continuous compounding with a shearing roll mill.

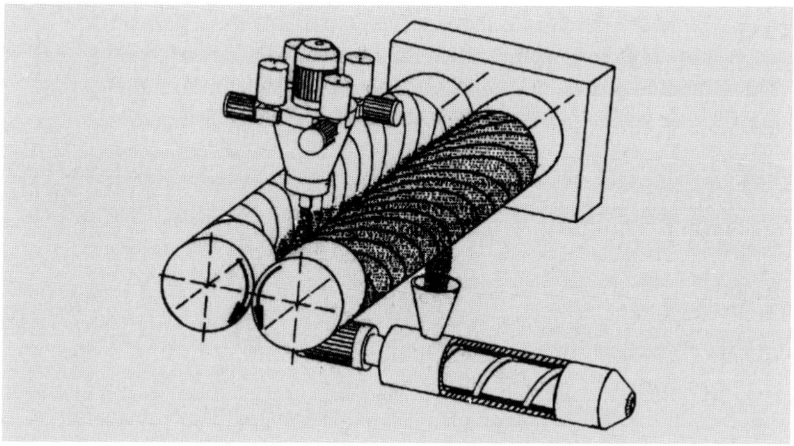

Figure 10-5 Continuous compounding with a shearing roll mill

internal mixer

The internal mixer is an example for a non-continuous compounding unit that compounds loads. It is also calaled ram kneader (Figure 10-6). It is especially

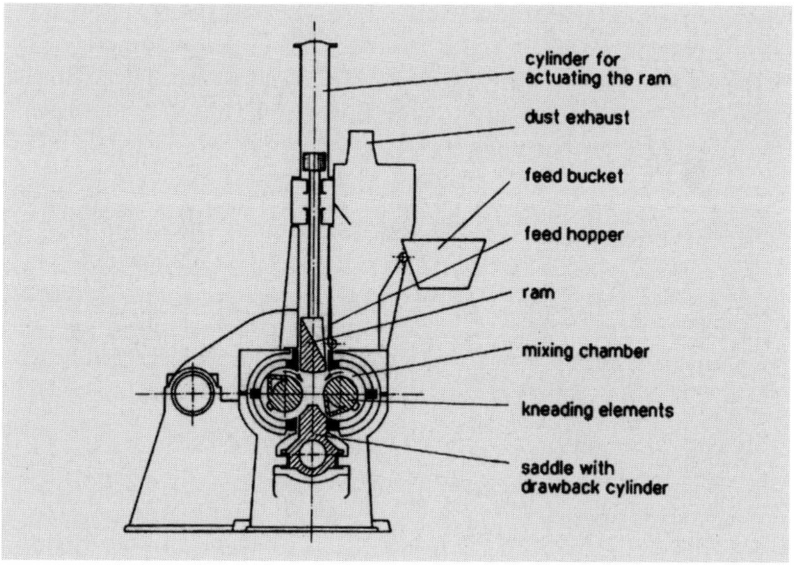

Figure 10-6 Internal mixer

well-suited to incorporating fillers, softeners, and chemicals in rubber mixtures and tough plastics. High shearing and elongation forces have to be used. A pneumatic or hydraulic ram closes the mixing chamber, which, however, will not be filled entirely.

10.5 Pelletizing

pelletizing methods

cold pelletizing

"Pelletizing" is the term used for the process of cutting the plastic into small, free-flowing pieces. Pelletizing methods include two variations, namely hot pelletizing and cold pelletizing methods. In cold pelletizing, the plasticized plastic is first cooled and then cut into pieces (see Figure 10-7). The disadvantage is that the cutting operation leaves ridges on the resulting pieces, which cause the pieces to become wedged easily, which in turn prevents them from flowing as freely as the pieces produced by hot pelletizing.

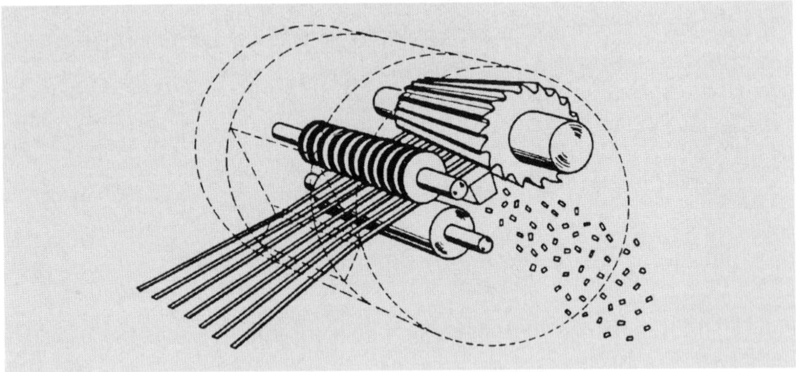

Figure 10-7 Strand pelletizer

hot pelletizing

In hot pelletizing, the plastic is plasticized in an extruder. The material is forced through a simple perforated plate, which serves as the extruder die. The emerging strands are cut by a knife, and the resulting pieces are cooled by air or water. This method is shown in Figure 10-8. An advantage of the method is that the pieces, while still warm, develop a shape free of ridges and sharp edges, which allows them to flow more freely.

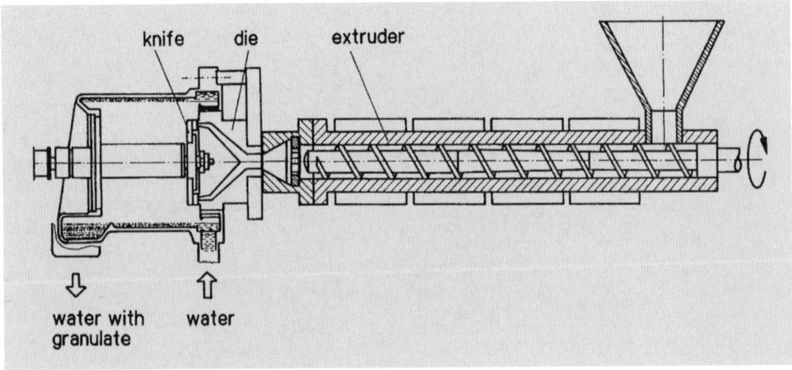

Figure 10-8 Hot pelletizing

10.6 Size Reduction

size reduction

Size reduction converts the plastic to a form that is easier to process. In the section on pelletizing we became acquainted with one application of size reduction. Recycling is another area becoming increasingly important. In recycling, pieces of scrap or collected plastic wastes are reduced in size and then reutilized. Cutting mills (Figure 10-9) are often used for this purpose.

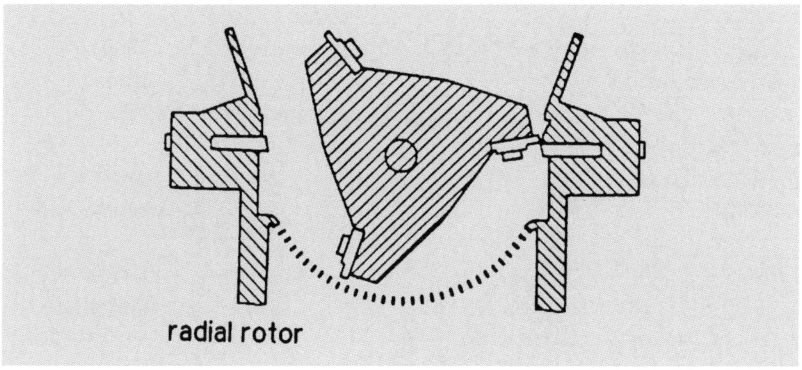

Figure 10-9 Cutting mill

Review Questions

No.	*Question*	*Choices*
1	Additives are added to plastics in order to improve their performance characteristics and ______________________ properties.	inspection distribution processing
2	______________________ are used to distribute the additives as evenly as possible throughout the plastic.	mixers kneaders mills
3	It is best to meter the additives according to ___________________ because this method is more precise.	weight volume
4	A kneader is used in the ______________________ of the plastic.	plasticizing mixing size reduction pelletizing
5	Hot-pelletized plastic flows ______________________ cold-pelletized plastic.	less freely than as freely as more freely than
6	In the recycling of plastic scrap or wastes, ____________________ are used to reduce the size of the pieces.	cutting mills strand pelletizers shearing roll mills

Lesson 11

Extrusion

Key Questions

What characterizes the extrusion process?
What components make up an extrusion line?
What functions are performed by the individual equipment components?
What products are manufactured by extrusion?

Subject Area

From Raw Plastic to Finished Product

Contents

Prerequisite Knowledge

Classification of Plastics (Lesson 5)
Physical Properties (Lesson 8)

11.1 Introduction

continuous semifinished product

Extrusion is the continuous manufacture of an "endless semifinished product" from plastic. The range of products extends from simple semifinished products, such as tubes, sheets, and films, to complicated profiles. It is also possible for the semifinished product to directly undergo further processing while it is still warm. This processing may include blow molding or calendering, for example. Extrusion is considered a primary processing method, because the plastic is completely fused and given an entirely new shape in this process.

primary processing method

11.2 Extrusion Equipment

Figure 11-1 is a diagram of an extrusion line. The structure and function of the individual components will be explained.

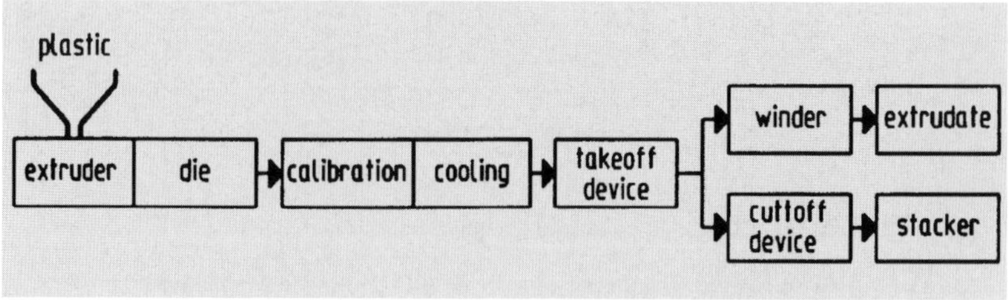

Figure 11-1 Extrusion plant

Extruder

homogeneous melt

The extruder is the component common to all extrusion lines and all methods based on extrusion. The extruder creates a homogeneous melt from the plastic (usually pellets or powder) fed to it. It then forces this melt through the die under the required amount of pressure. An extruder consists of the components shown in Figure 11-2.

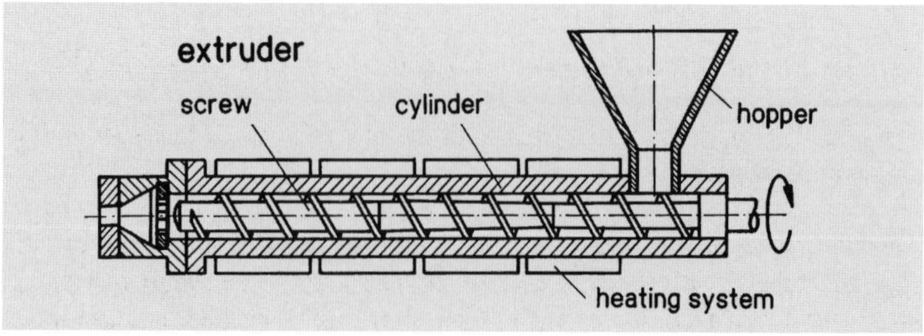

Figure 11-2 Extruder

Hopper

The hopper provides the extruder with a steady feed of the material being processed. Not all materials tend to flow freely, however, so the hopper is often equipped with a supplementary agitator or conveyor.

feed

Screw

The screw performs a multitude of functions, including the feeding, conveyance, fusion, and homogenization of the plastic. It is therefore the heart of the extruder. The most widely used type is the "three-zone screw" (Figure 11-3), because it can process most thermoplastics satisfactorily with regard to thermal and economic aspects. For this reason, we will consider the three-zone screw to be representative of all types of screws.

three-zone screw

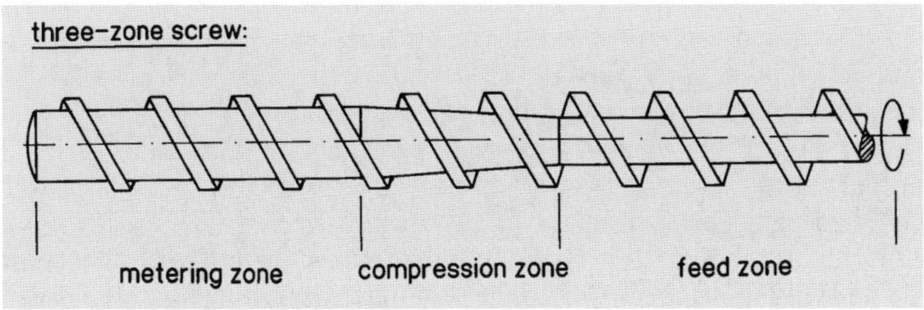

Figure 11-3 Three-zone screw

In the feed zone, the solid material is drawn in and then forced forward. In the compression zone, the material is compressed and fused by the decreasing flight depth of the screw. In the metering zone, the fused material is homogenized and brought to the desired processing temperature.

feed zone
compression zone
metering zone

One important characteristic of the screw is the ratio of its length to its outer diameter, the L/D ratio. This ratio determines the efficiency of the extruder.

L/D ratio

In addition to the commonly used three-zone screw, other screw forms are used for special applications. Regardless of their design, however, all screws, and thus all extruders, must satisfy the following requirements:

extruder requirements

1. Constant transport, virtually free of pulsations,
2. production of a thermally and mechanically homogeneous melt,
3. processing of the material below its thermal, chemical, and mechanical degradation limits

From the standpoint of economy, there is also a demand for a high rate of material output at a low specific operating cost. But these requirements can only be fulfilled when the screw and barrel are well matched, because these two components work closely together.

Barrel

barrel design

Individual extruders are distinguished by barrel design (see Figure 11-4).

extruder	**barrel design**
single-screw extruder	conventional grooved
twin-screw extruder	co-rotating counter-rotating

Figure 11-4 Classification of extruders by barrel design

single-screw extruder, conventional

The conventional single-screw extruder has a barrel with a smooth inner surface. It is characterized by the fact that the pressure needed for overcoming the die resistance is developed in the metering zone. The feed material is transported by the solid friction that develops between the material particles themselves and between the particles and the barrel wall.

single-screw extruder, with grooved barrel

In the grooved-barrel single-screw extruder, the barrel wall is provided with tapered longitudinal grooves in the feed zone. These grooves increase the transport, and thus the compression, of the material. In this type of extruder, the buildup of pressure begins in the feed zone. Special mixing components must be used in the metering zone, however, because it has been found that this type of extruder does not homogenize the material as well as the conventional design.

twin-screw extruder, counter-rotating

The counter-rotating twin-screw extruder is used for powdered materials, especially PVC. An advantage of this extruder is that it makes it easier to mix additives into the plastic without subjecting the material to intense mechanical or thermal stress.

In the barrel, which has a cross section resembling a figure eight, the screws are arranged to form closed chambers between the flights. The material is force-fed in these chambers (see Figure 11-5). Near the end of the screws (where the pressure builds), a leakage flow finally develops, and the material melts by friction.

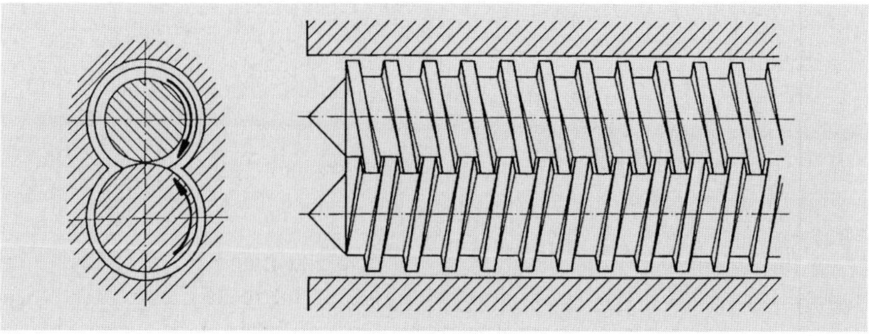

Figure 11-5 Twin-screw extruder, intermeshing counter-rotating

The advantage of the twin-screw extruder is that it allows sensitive materials to be processed at low residence times and high temperatures without exceeding degradation limits.

twin-screw extruder, co-rotating

The co-rotating twin-screw extruder is used primarily for compounding. It conveys the material by means of frictional contact between the screws and the barrel. The screws are so designed as to keep each other clean as they rotate.

Heating System

introduction of heat

The fusion of the material in the extruder is performed not only by friction but also by the introduction of heat from external sources. This is the task of the heating system. The system is divided into multiple zones that can be heated or cooled separately. Band heaters are most commonly used for this purpose, but other systems (e.g., fluid circulation systems) may also be used. It is thus possible to achieve a certain temperature distribution along the barrel. Heated screws are also used to process some thermally sensitive materials.

Processed Materials

viscosity differences

The extrusion process is used for the same types of plastics that can also be processed by injection molding. However, there is a big difference between the two processes. As a result, different demands are placed on the materials being processed. While the characteristics considered desirable for injection molding include low viscosity and high flowability, extrusion requires high viscosity. From the time the material leaves the die until it enters the calibrator, this high viscosity ensures that the material maintains its form and does not flow off. Figure 11-6 lists several examples of applications (extrudates)—in other words, products manufactured by the extrusion method.

plastics	**range of processing temperatures**	**application examples (extrudates)**
PE	130–200°C (205–390°F)	tubes, sheets, films, coverings
PP	180–260°C (355–500°F)	tubes, flat films, sheets, tapes
PVC	180–210°C (355–410°F)	tubes, profiles, sheets
PMMA	160–190°C (320–375°F)	tubes, profiles, sheets
PC	300–340°C (570–645°F)	sheets, profiles, hollow bodies

Figure 11-6 Typical extrudates

Principle of Extruder Function

extruder principle

The principle underlying the extruder's function is similar to that of a meat grinder. As mentioned, the feed zone draws the material in and then passes it on to the compression zone. There it is compressed by the decreasing flight and increasing diameter of the root of the screw, desecrated (in some cases), and converted to the molten state. In the subsequent metering zone, the material is further homogenized and uniformly heated (see Figure 11-3).

mixing zones

The pressure may develop in the feed zone or metering zone, depending on the type of extruder. Because the melting process does not always produce a completely fused, homogeneous melt, mixing zones are built into the screw in such cases (see Figure 11-7).

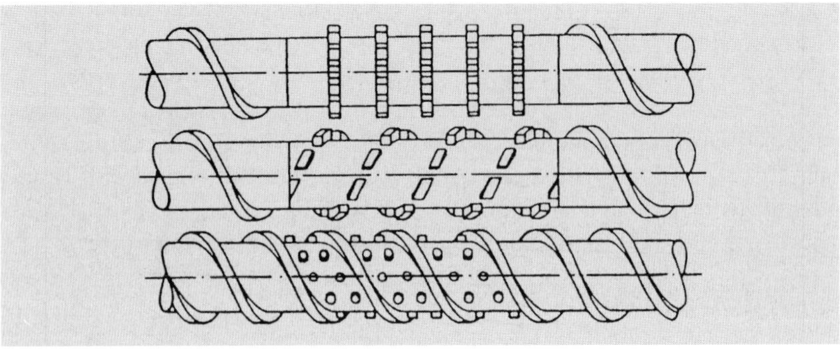

Figure 11-7 Mixing zones

Dies

dies

The extruder converts the material into a homogeneous melt, and the die, which is flange-mounted to the extruder, determines the shape of the extruded semifinished product (also known as "extrudate"). Various extrudates are distinguished by shape (see Figure 11-8).

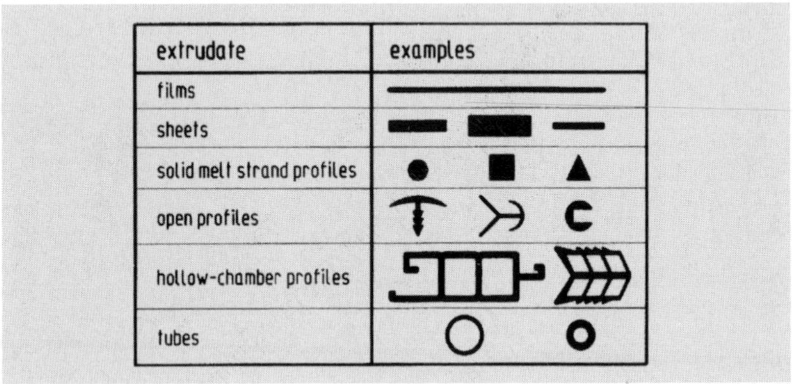
extrudate
examples
films
sheets
solid melt strand profiles
open profiles
hollow-chamber profiles
tubes

Figure 11-8 Shapes of various extrudates

manifold

All dies contain a flow channel, the "manifold", through which the melt flows. The manifold gives the melt its desired form. As a rule, all dies are electrically heated. Several dies will be discussed.

Displacement- or Spider Dies

Spider dies are predominantly used for the manufacture of rigid and flexible tubes and tubular films (see Figure 11-9). These dies feature a displacer, shaped to assist the flow as much as possible, connected to the outer wall of the manifold by fins. It has a conic shape on the side facing the extruder. Towards the discharge end, it takes on the desired inner shape of the extrudate. The advantage of the spider die consists in the fact that the oncoming flow is well centered. This results in a favorable distribution of the melt. The spider fins have a disadvantageous effect, however. As the melt flows around the fins, the fins create flow marks. These marks manifest themselves as localized weak spots and streaks in the semifinished product.

spider die

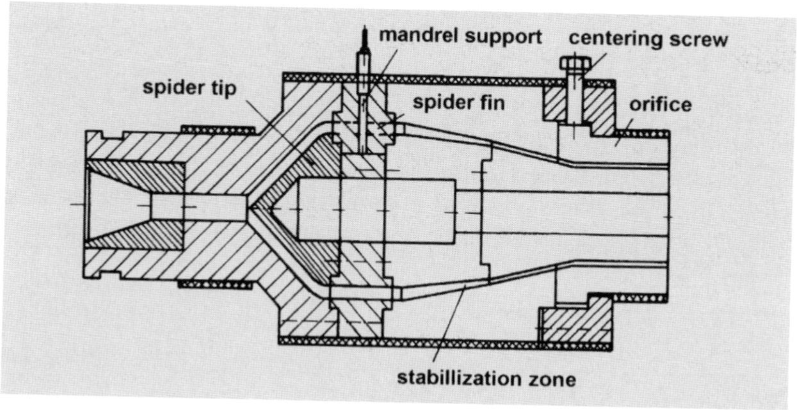

Figure 11-9 Spider die

In order to avoid such flow marks, slurring threads or spiral mandrel distributor dies are used (see Figure 11-10). The function of the slurring threads is to overlap the axial flow with a tangential component, thus distributing the flow marks evenly across the semifinished product.

slurring threads

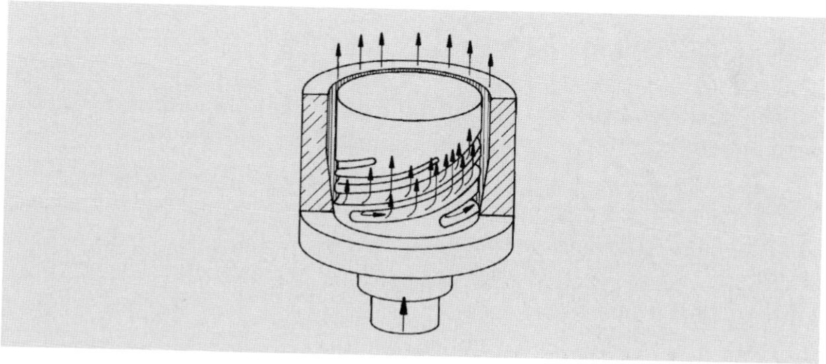

Figure 11-10 Spiral mandrel distributor die

The spiral mandrel distributor die does not contain a mandrel support element. The initially radial flow is converted to an axial flow by this die.

spiral mandrel distributor die

Slit Distributor Dies

slit distributor die

Slit distributor dies are used to manufacture flat films and sheets (see Figure 11-11).

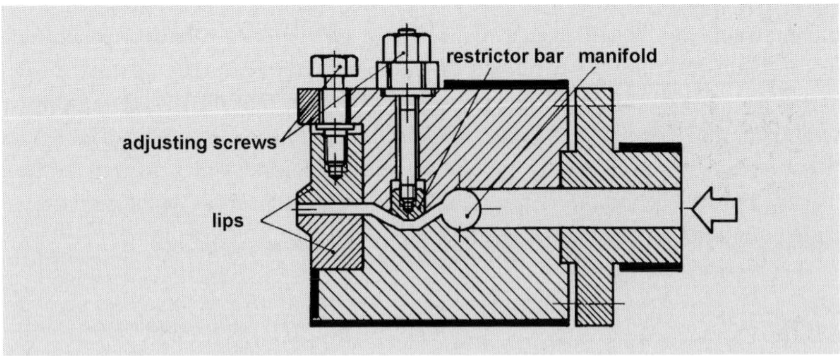

Figure 11-11 Slit distributor die

coathanger die

Slit distributor dies first spread the flow of melt to a greater width and then shape it into a thin layer. The melt strand, which is usually round, initially enters a manifold, where it spreads to form a rectangular melt web. In most cases, this manifold takes the shape of a coathanger (see Figure 11-12).

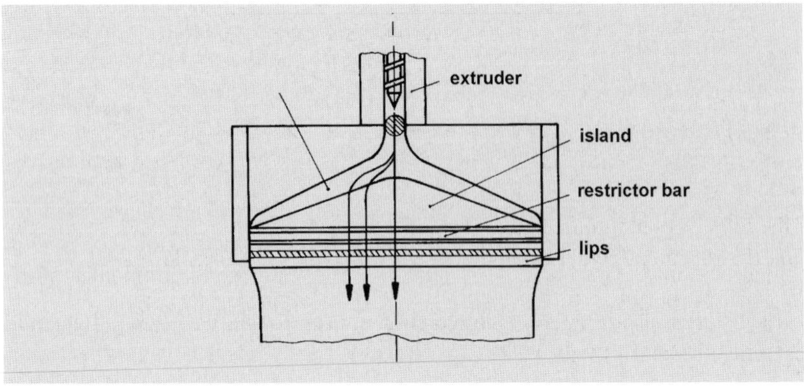

Figure 11-12 Coathanger die

island region

The melt then proceeds into the "island region" with the restrictor bar. The island region leads to the lips, where the melt flows out of the die.

In addition to those presented here, many other dies are used for special purposes (for example, cable coatings).

Downstream Components

calibrator

After the melt leaves the extruder die, the shape and dimensions of the melt must be fixed. This is the job of the calibrator, which functions by means of compressed air or vacuum. The extrudate is pressed against the wall of the calibrator and cooled sufficiently to prevent its further deformation in the subsequent cooling section.

cooling section

The lengths of the calibrating and cooling sections must be matched to the throughput of the extruder, and their shapes must be matched to the desired shape of the extrudate. Although flat extrudates are cooled by rolling, the extrudates of profiles, tubes, cables, and similar shapes are passed through water baths. Air-cooling systems and water-spray-cooling systems are also commonly used.

takeoff device

A takeoff device follows the cooling section. Its function is to draw the extrudate from the die through the calibrating and cooling sections at a constant rate. The extrudate is able to withstand the considerable drawing forces without suffering deformation, because it has previously been strengthened in the calibrating and cooling sections.

cutoff device

The final station in an extrusion line is formed by the cutoff and stacking device for rigid tubes, sheets, and profiles, or by the winding device for films, cables, filaments, and flexible tubes.

11.3 Coextrusion

explanation

Coextrusion is used when the requirements placed on the extrudate cannot be fulfilled by a single material, or when material costs can be reduced by combining two highly stress-resistant outer layers with an economical inner layer. The semifinished product is then manufactured from multiple layers of different materials.

coextrusion die

In order to manufacture such a composite of different materials, each material is plasticized in a separate extruder. In a special coextrusion die (Figure 11-13), each of the different melts is formed in a separate manifold. The melts are combined and fused just before leaving the die. It is now possible to manufacture composites with as many as seven layers.

Coextrusion is now used for multilayer cable insulation, packaging films, and extrusion blow molding.

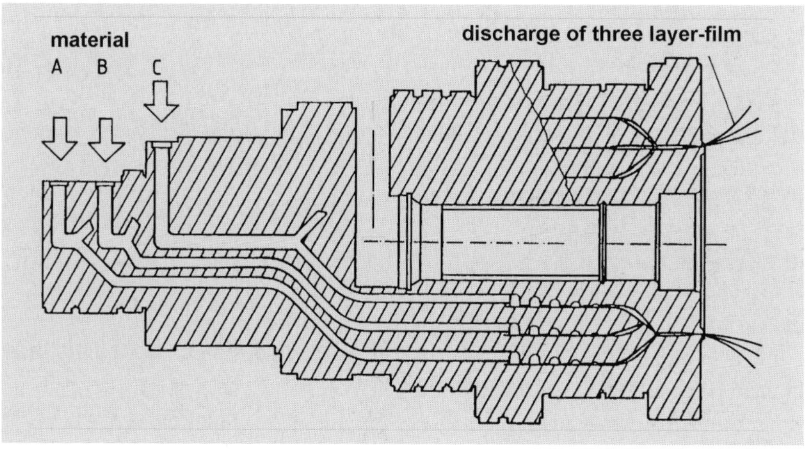

Figure 11-13 Three-layer melt manifold

11.4 Extrusion Blow Molding

products

Extrusion blow molding is a modern method of producing hollow-bodied products (e.g., motor vehicle fuel tanks, canisters, surfboards, heating oil tanks, bottles) from thermoplastics. This technique requires two major equipment components:

equipment components

1. An extruder (usually a single-screw extruder) with a cross head,
2. the blow mold and blowing station

process sequence

The process sequence for extrusion-blow molding is represented in Figure 11-14.

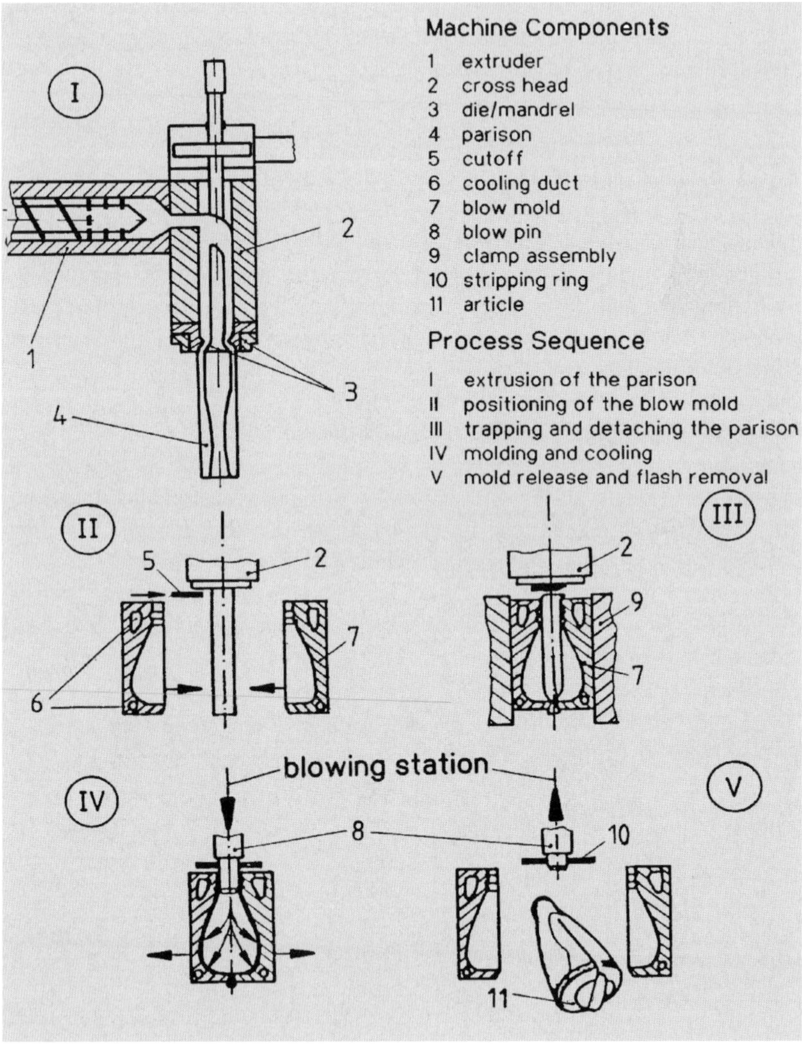

Figure 11-14 Extrusion blow molding

parison

The extruder processes the plastic to form a homogeneous melt, as described. As the melt exits the horizontal extruder, the attached cross head turns it to the vertical direction. A die subsequently shapes the melt into a tubular parison. This parison then hangs freely.

The blow mold consists of two movable halves that display the opposite shape of the part being manufactured. After the parison has exited the cross head, the mold closes around it and pinches it shut at the lower end. The mold is then moved from the machine frame to the blowing station. At the blowing station, the blow pin drops into the mold (and thus into the parison). In this operation, the blow pin shapes and calibrates the neck region of the hollow body. At the same time, compressed air is blown into the parison. (See Figure 11-15).

blow mold

blow molding

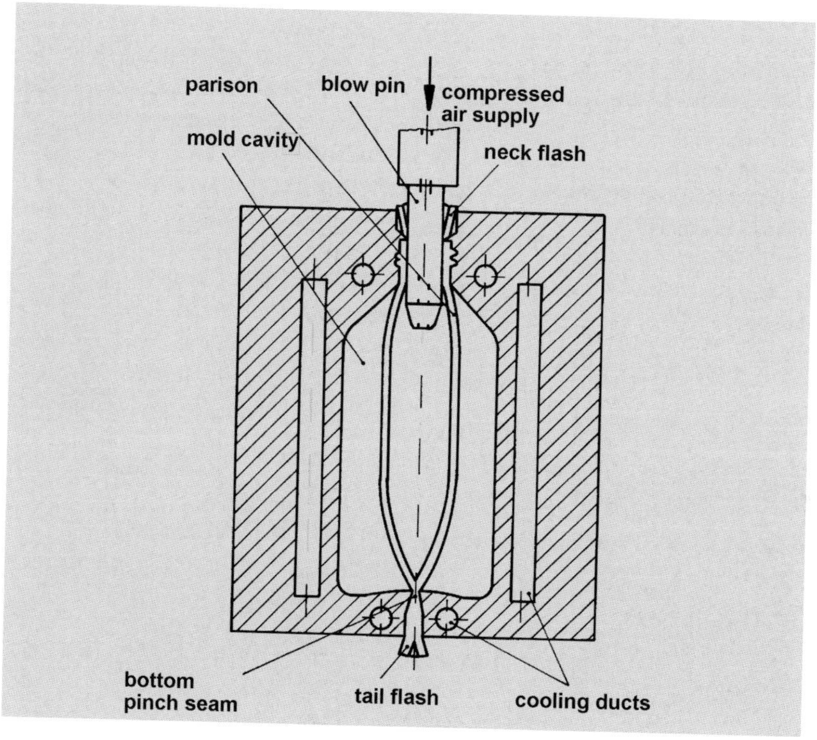

Figure 11-15 Blow mold

As a result of the pressure created by the compressed air within the parison, the parison is inflated and pressed against the mold walls. In this way, the parison is given the desired shape. At this moment, the mold also begins to perform its cooling function.

compressed air

In order to shorten the cooling period, we must create air circulation within the mold component, by bringing an exhaust bore in contact with the blow pin. The air can thus be discharged through a throttle that serves to maintain the desired blowing pressure. Aside from air with CO_2, cooled nitrogen can also be used as the blowing medium.

cooling

After the molded part has been adequately cooled, and thus adequately strengthened, the blow molding head is retracted, the mold is opened, and the molded part can be removed.

moded part removal

Review Questions

No.	*Question*	*Choices*
1	In the extrusion process, products are manufactured ___________.	continuously batchwise
2	Which component in an extrusion line is responsible for melting the plastic homogeneously? ___________________________________	the calibrator the extruder the die
3	The most commonly used type of screw is the _________________.	devolatilizing screw short compression screw three-zone screw
4	In order to prevent the extrudate from "flowing off" as it leaves the die, the processed plastic should display _____________________ viscosity.	low high
5	The die determines the _______________________of the extrudate.	length shape temperature
6	Coextrusion is used to manufacture films and sheets with ______________________.	a single layer multiple layers
7	Motor vehicle fuel tanks and bottles are manufactured by ______________________.	coextrusion extrusion blow molding

Lesson 12

Injection Molding

Key Questions

How is an injection molding machine constructed?
What are the functions of the individual components?
What is the sequence of steps in the injection molding process?

Subject Area

From Raw Plastic to Finished Product

Contents

12.1 Introduction
12.2 Injection Molding Machine
12.3 Mold
12.4 Process Sequence
12.5 Injection Molding of Thermosets and Elastomers

Review questions

Prerequisite Knowledge

Classification of Plastics (Lesson 5)

12.1 Introduction

injection molding

Injection molding represents the most important process for the manufacture of molded parts from plastic. About 60% of all plastics processing machines are injection molding types. These machines make it possible to manufacture molded parts weighing anywhere from a few mg to 90 kg (less than an ounce to 200 lb). Injection molding is considered one of the primary processing methods. The injection molding process is represented schematically in Figure 12-1.

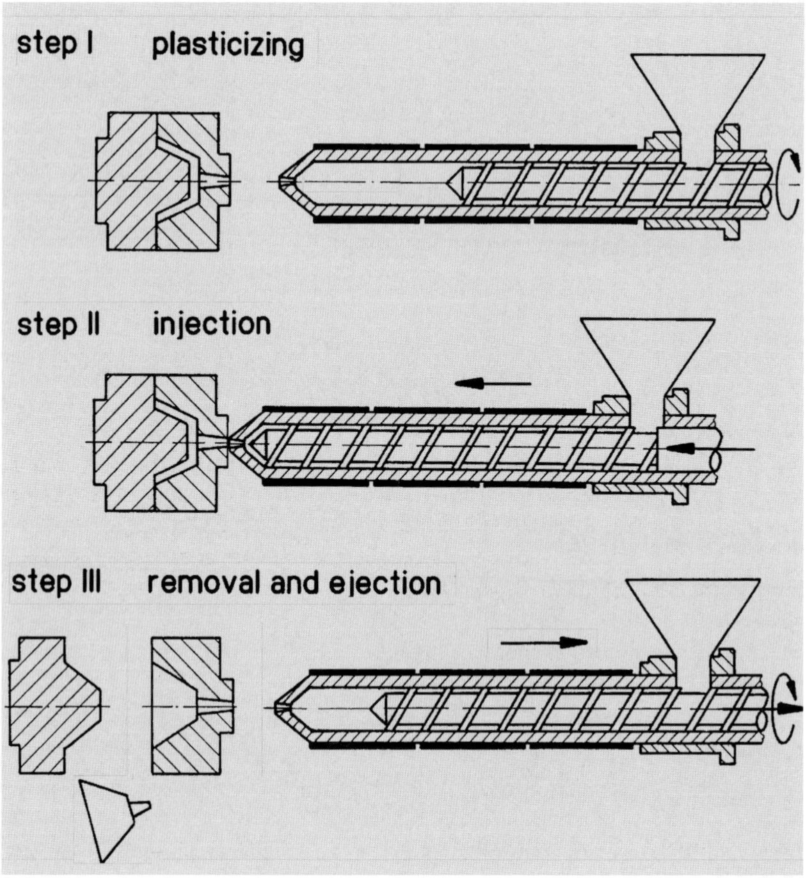

Figure 12-1 Injection molding process (schematic)

mass-produced goods

minimal finishing

Injection molding is suitable for mass-produced goods, because the raw material usually can be converted to a finished part in only one operation. In contrast to metal casting or the compression molding of thermosets and elastomers, injection molding of thermoplastics produces no flash (provided that mold quality is high). The need for finishing the injection-molded part is therefore minimized or completely eliminated. It is thus possible to produce even complicated geometrical shapes in a single operation.

Most plastics processed by injection molding are thermoplastics, but thermosets and elastomers can also be injection molded (see Figure 12-2).

thermoplastics	thermosets	elastomers
polystyrene (PS)	unsaturated polyester resin (UP)	nitrile rubber (NBR)
acrylonitrile/butadiene/ styrene (ABS)	phenol-formaldehyde resin (PF)	styrene/butadiene rubber (SBR)
polyethylene (PE)		polyisoprene (IR)
polypropylene (PP)		
polycarbonate (PC)		
polymethyl methacrylate (PMMA)		
polyamides (PA)		

Figure 12-2 Plastics for injection molding

The profitability of injection molding is decisively influenced by the output of parts per unit time. This output rate depends heavily on the length of time required for the molded part to cool within the mold. The cooling time in turn depends on the greatest wall thickness of the molded part. The cooling time increases as a square of the wall thickness! This has to be considered for molded parts with great wall thickness and is economically very important. The time that elapses between the discharge of two consecutive finished molded parts is called the "cycle time."

cooling time

cycle time

The features of injection molding include:

features

— Direct path from molding compound to finished part,

— requires little or no finishing of molded part,

— process can be fully automated,

— high reproducibility of molded parts,

— high quality of molded parts.

12.2 Injection Molding Machine

Most injection molding machines are general purpose machines. The tasks they perform include the batch manufacture of molded parts from molding compounds. Shaping of parts occurs under pressure. These tasks are fulfilled by the various subassemblies that make up an injection molding machine (see Figure 12-3).

definition

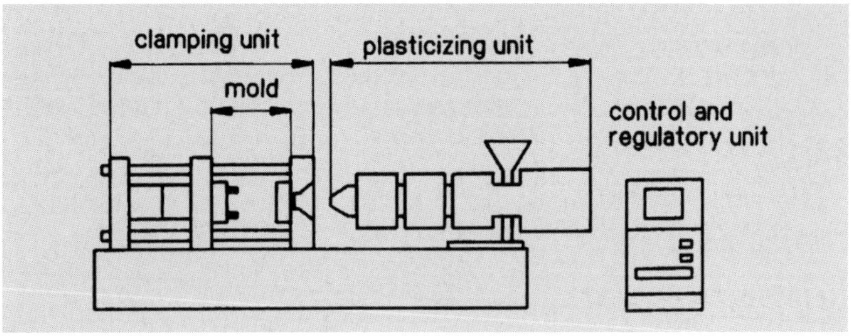

Figure 12-3 Structure of an injection molding machine

Injection Unit

functions

The injection unit melts the plastic, homogenizes it, conveys it, meters it, and injects it into the mold. The injection unit thus has two overall tasks: the plasticizing of the plastic, and the injection of the plastic into the mold. The use of machines with reciprocating screws is now commonplace. These injection molding machines work with a screw that also serves as an injection ram (see Figure 12-4).

The screw turns within a cylinder that can be heated. The material is fed into this cylinder through a hopper from above.

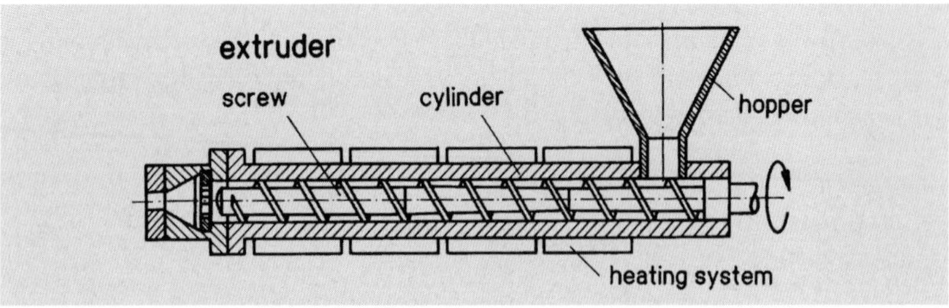

Figure 12-4 Injection unit of a screw injection molding machine

The injection unit is generally mounted on the machine bed in such a way that it can be moved. As a rule, it is possible to change cylinders, screws, and nozzles in order to adapt to the substance being processed or the shot volume.

Clamping Unit

The clamping unit of an injection molding machine is comparable to a horizontal press. The nozzle platen is fixed, and the design of the clamp platen allows it to slide on four tie rods. The molds are attached to these vertical platens in a manner that allows the finished molded parts to fall out at the bottom.

drive systems

The two most commonly used drive systems for the clamp platen are:

1. Hydraulically operated toggle,
2. fully hydraulic clamping unit.

toggle systems

Toggle systems are used in small to medium-sized machines. The toggle is hydraulically driven (see Figure 12-5).

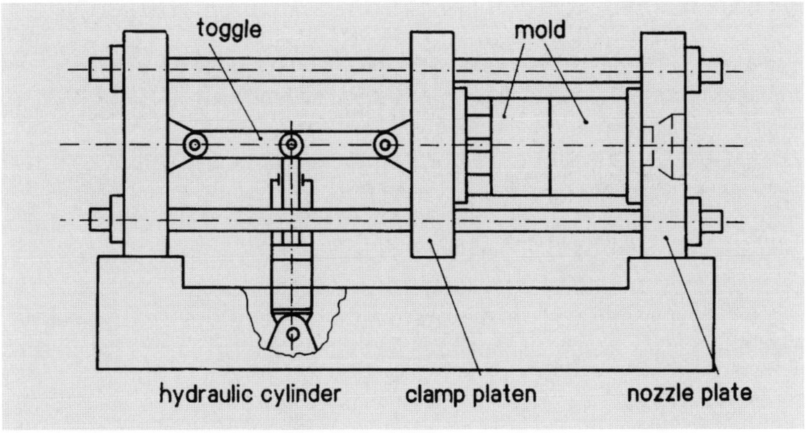

Figure 12-5 Toggle clamp unit

The advantages of this system are its self-locking operation and the quick and favorable process of motion and speed. The disadvantages are possible tie rod fractures and permanent deformation of the mold, which may occur when the system is poorly adjusted, and the high cost of maintenance.

hydraulic clamping unit

The danger of tie rod fractures does not exist in the fully-hydraulic system (Figure 12-6), because the hydraulic fluid can yield and thus absorb excessively severe deformations.

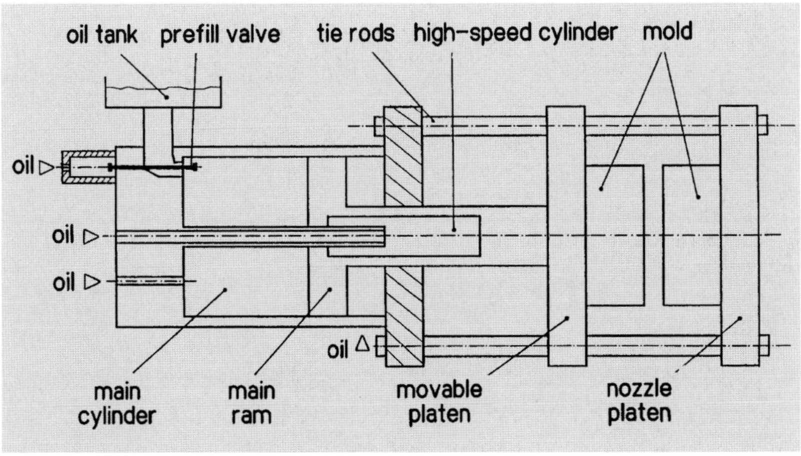

Figure 12-6 Fully-hydraulic clamping unit

The advantages of this system are its greater precision, variable positioning, and freedom from the danger of excessive mold deformation and tie rod fractures. The disadvantages of the system include its lower clamping speed and the lower rigidity of the clamping unit, which is caused by the strong tendency of the oil to yield. Another disadvantage is the increased energy requirement.

Machine Bed and Control Cabinet

machine bed

The machine bed holds the plasticizing and clamping units, the hydraulic fluid container, and the hydraulic drive. The control and regulatory apparatus is also housed directly within the machine bed in some cases.

control cabinet

The control cabinet contains the instruments, electrical controls, regulators, and power supply system. This represents the unit that controls and/or regulates the machine. In modern machines, the input of parameters is performed by means of keyboards and interactive screen programs. The microcomputer mounted in the control cabinet provides process control, monitors process and production data, saves data, and documents the process.

12.3 Mold

injection mold

The mold does not belong to the injection molding machine itself, because it must be specially designed for each individual molded part. The mold consists of at least two main components, each of which is fastened to one of the platens of the clamping unit. The maximum mold size is determined by the size of the platens and the distance between two adjacent tie rods of the injection molding machine.

mold components

The injection mold essentially consists of:

— Platens with cavity,

— gating system,

— heating system,

— ejection system.

functions

These elements essentially perform the following functions:

— Receiving and distributing the melt,

— forming the melt into the shape of the molded part,

— cooling the melt (thermoplastics) or supplying the activation energy (elastomers and thermosets),

— removing the part from the mold.

classification criteria

Figure 12-7 shows an example of an injection mold.

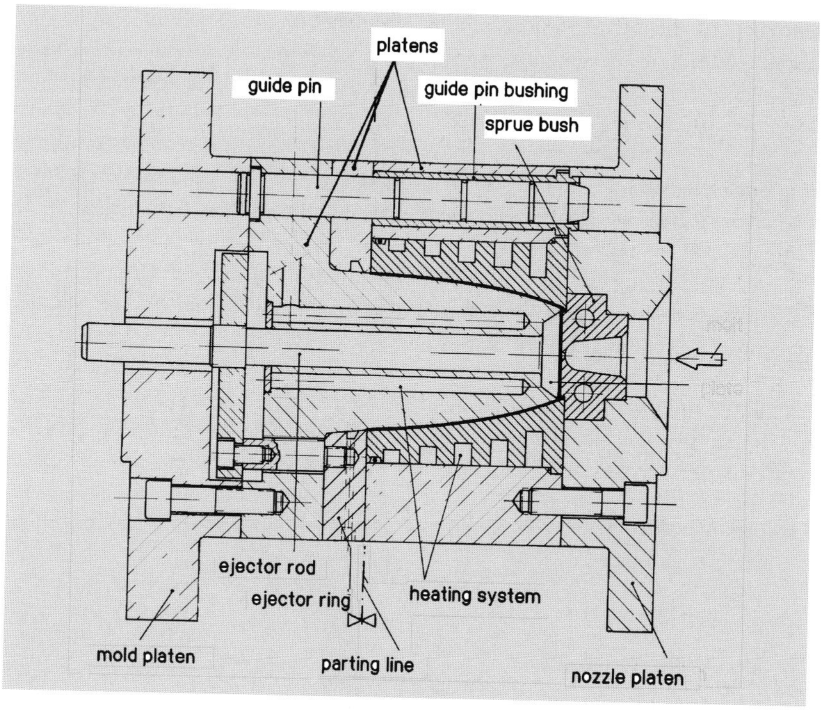

Figure 12-7 Injection mold

Molds are classified by the following criteria:

— Basic structure,

— type of removal system,

— type of gating system,

— number of cavities,

— number of parting lines,

— size of mold.

The cost of molds is very high. In general, molds cost from 10,000 to a few hundred thousand dollars. For this reason, their purchase is worthwhile only for large-scale production. *mold cost*

Removal System

One of the functional elements that can be moved is the ejector unit, with its ejector plates and pins. At the end of the cooling period, the mold is opened by the clamping unit. The ejector bolts are moved toward the molded part by a hydraulic cylinder. The molded part is pushed out of the cavity by the ejector pins and then falls out of the mold. *ejector unit functions*

During the injection phase, the melt is pushed through the gating system and conveyed through the gate and into the cavity. The cavity shapes the molded part. The gating system can be designed as a heated or unheated system.

multicavity molds

Greater economy can be attained through the use of multiple cavities within a single mold. In this case, the gate is constructed in such a way that a channel leads from the direction of the plasticizing unit and then branches into multiple channels. These multiple channels, or "runners," lead in turn to the individual cavities. The design of the gating system should allow the individual cavities to be filled simultaneously, and the melt entering the different cavities displays the same temperature and pressure.

12.4 Process Sequence

injection molding cycle

The production sequence, generally known as the "injection molding cycle," can be seen in Figure 12-8.

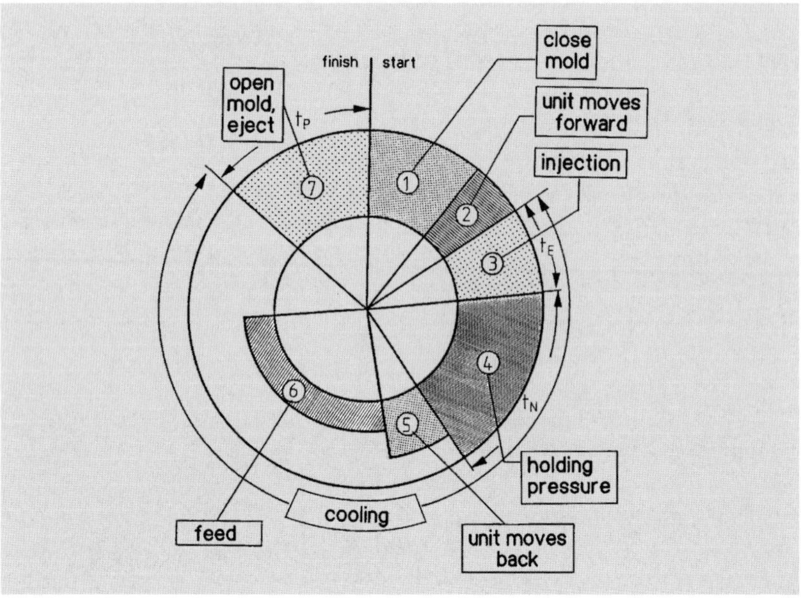

Figure 12-8 Injection molding cycle

chronological sequence

In order to clarify the chronological sequence of the individual process steps, the operations are presented schematically over a time line in Figure 12-9.

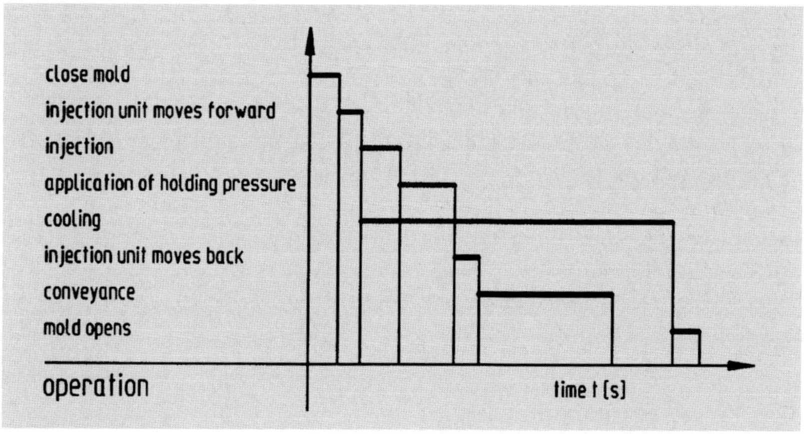

Figure 12-9 Chronological sequence of steps in an injection molding cycle

This clearly shows that the process steps occur consecutively, except for the important cooling process, which overlaps other processes. These process steps are coordinated by the machine's control apparatus and repeated in each injection cycle. The cycle time should be kept as low as possible in order to attain a high output performance and, thus, high profitability.

profitability

Feed (Metering)

The material is transported from the hopper toward the screw tip by means of a screw. This screw turns within a cylinder and compresses and melts the material. At the same time that the screw transports material, it is pressed backward by the material that accumulates in front of the screw tip. Transport of the material stops when the screw has reached a certain position (see Figure 12-10).

screw tip

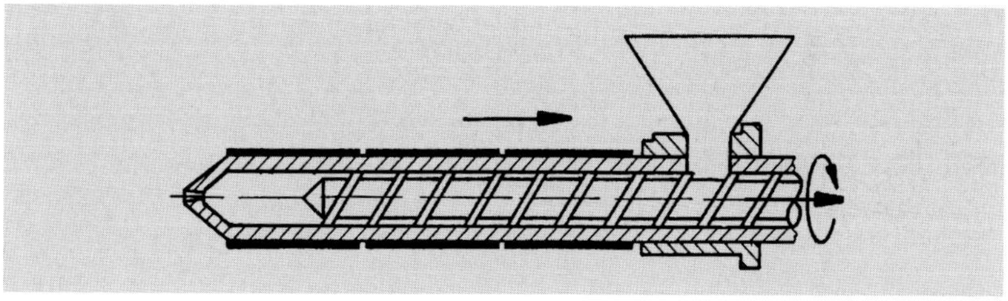

Figure 12-10 Screw position after feeding

At this point, a sufficient amount of material has accumulated in front of the screw tip to allow the molded part to be injected. The distance that the screw has traveled is called the "feed distance." The volume of material in front of the screw is called the "feed capacity." Both parameters are specially matched to each mold.

feed distance
feed capacity

Injection

back-flow valve

ram

In the injection phase, the hydraulic injection cylinder drives the screw forward without causing it to rotate. The screw pushes the metered melt through a nozzle and into the mold. The screw functions as a ram in this phase. This function is controlled by the back-flow valve.

injection pressure

injection rate

The injection pressure is specified on the machine as a fixed parameter. It represents an upper limit that may not be exceeded. Another parameter that must be set on the machine is the injection rate, which, however, can be adjusted during injection.

Flow Processes within the Mold

phases

The flow processes within the mold can be divided into three phases:

Phase I: Injection phase

Phase II: Compression phase

Phase III: Holding pressure phase

injection phase
compression phase

The volume of the mold is filled during the injection phase. As soon as the mold is full, the melt begins to slow down. The compression phase is now beginning. In order to compress the molded part, additional molding compound (an additional 7% or so) is conveyed into the mold. The pressure within the cavity increases sharply during the compression phase. When a specified pressure level has been attained within the mold, the transition to the holding pressure phase occurs.

shrinkage

holding pressure phase

The material shrinks while it cools in the cavity. A feed of additional material must therefore be provided, in order to maintain the molded part at a constant volume. This is the purpose of the holding pressure phase. The pressure within the molded part decreases over time—even when the holding pressure is maintained at a constant level—because the molded part becomes increasingly solid. When the pressure within the molded part has fallen to ambient level, the holding pressure phase is over.

transition point

The time at which the transition to the holding pressure phase occurs is important. If the transition is made too early, the molded part is insufficiently compressed and sink marks result. If the transition is made too late, it can cause excessive injection and thus the formation of flash on the molded part. After the holding pressure phase has ended, the injection unit immediately begins the next feed operation.

Cooling Process

cooling period

The cooling period begins with the filling process and ends with removal of the part from the mold. This period is adjusted to bring the molded part to a certain final temperature, thus stabilizing its form. The cooling process is

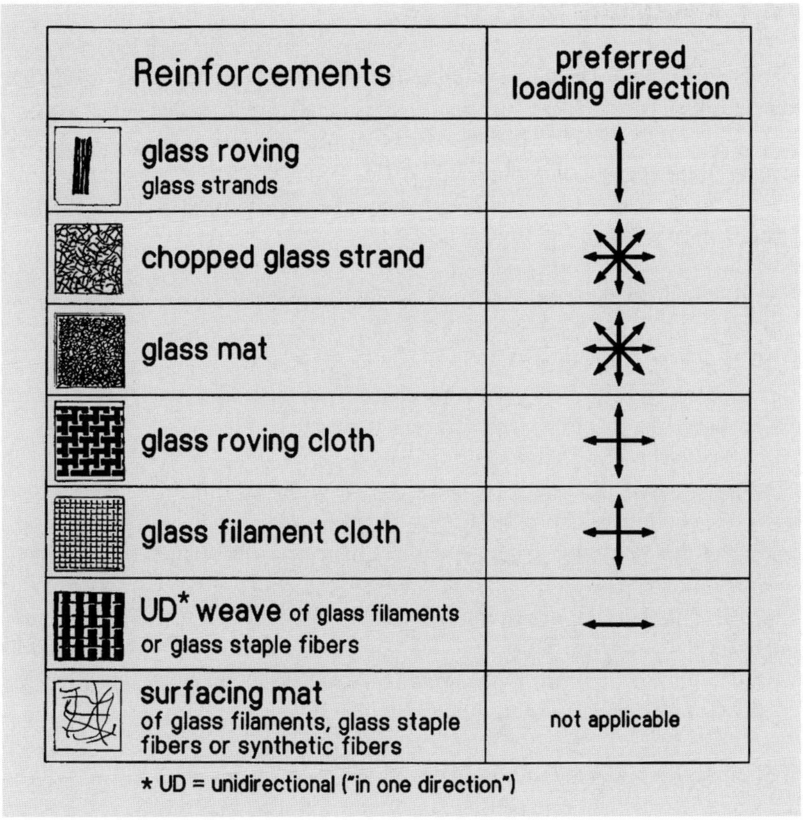
Reinforcements
preferred loading direction
glass roving
glass strands
chopped glass strand
glass mat
glass roving cloth
glass filament cloth
UD* weave of glass filaments
or glass staple fibers
surfacing mat
of glass filaments, glass staple fibers or synthetic fibers
not applicable
* UD = unidirectional ("in one direction")

Figure 13-2 Fiber forms

loading direction

The choice of the fiber form is determined by the manner in which the part will be loaded. For example, glass rovings can be heavily loaded in only one direction, whereas a glass roving cloth can be heavily loaded in two directions.

13.2 Process Sequence

process steps

The manufacture of parts from fiber-reinforced plastics generally occurs in four steps:

Step I: Applying and aligning the fibers

Step II: Impregnating the fibers

Step III: Shaping the part

Step IV: Curing the plastic
The sequence of the first two steps can be reversed. The sequence of steps differs from one manufacturing process to another.

thermoset matrix

Thermosets form the matrices of most fiber-reinforced plastics. As the part cures, the thermoset is produced by a chemical reaction involving the cross-linking of the resin with which the fibers have previously been impregnated.

hardeners
accelerators

In order to initiate the cross-linking reaction at room temperature, hardeners and/or accelerators are admixed with the resin. The use of these additives depends on the type of resin involved. Once the curing process has ended, the structure of the plastic can no longer be altered, even by heating.

air bubbles

The finished product should not be damaged by the forces to which it will be subjected in practical application. However, this can be ensured only if the fibers adhere very tightly to the plastic. This adherence can be impaired by air bubbles on the fibers. The fibers must therefore be free of any clinging air bubbles during the final curing of the plastic. If any air bubbles remain, the plastic could become detached from the fibers under heavy loading, thus destroying the part. When the fibers are impregnated with resin, it is thus important to ensure that no air bubbles enter the resin. Otherwise, the component will have to be compressed and deaerated before and/or during the curing process.

13.3 Manual Processing Methods

Hand Lay-Up

The simplest process for manufacturing components from fiber-reinforced plastics is called "hand lay-up." Alternating layers of resin and fiber mats are applied to a positive mold. The fiber mat is pressed on to the mold with a laminating roller, thus becoming very thoroughly impregnated with resin. This is shown in Figure 13-3.

positive mold

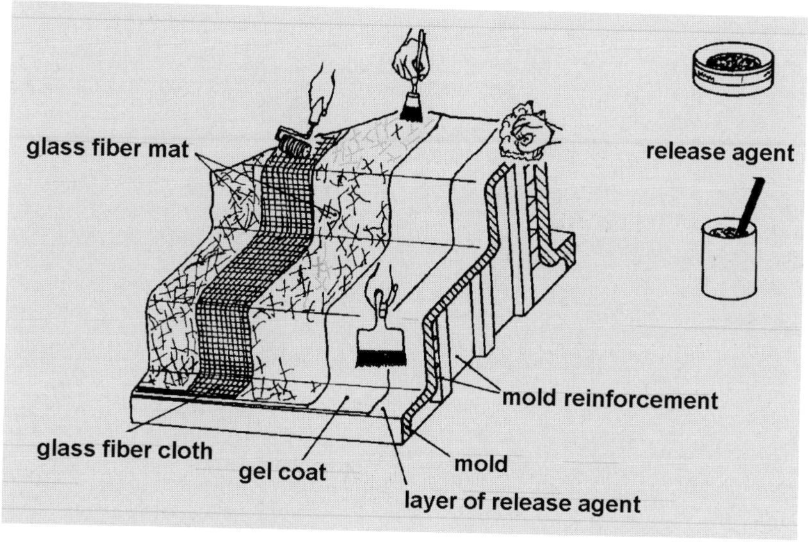

Figure 13-3 Hand lay-up

Before the actual laminating takes place, a release agent and a gel coat are applied to the mold. The release agent improves the separation of the finished component from the mold. The gel coat improves the surface of the molded part, insofar as the fibers are unable to penetrate this layer. One area where this process is used is boat construction.

surface application

13.4 Automated Processing Methods

Fiber Spray-Gun Molding

process

In fiber spray-gun molding, the chopped fibers are blown onto a mold by compressed air. The resin is simultaneously sprayed onto the mold from another nozzle. The resulting layer is compressed and simultaneously deaerated before the part cures. The fiber spray-gun molding process is represented in Figure 13-4.

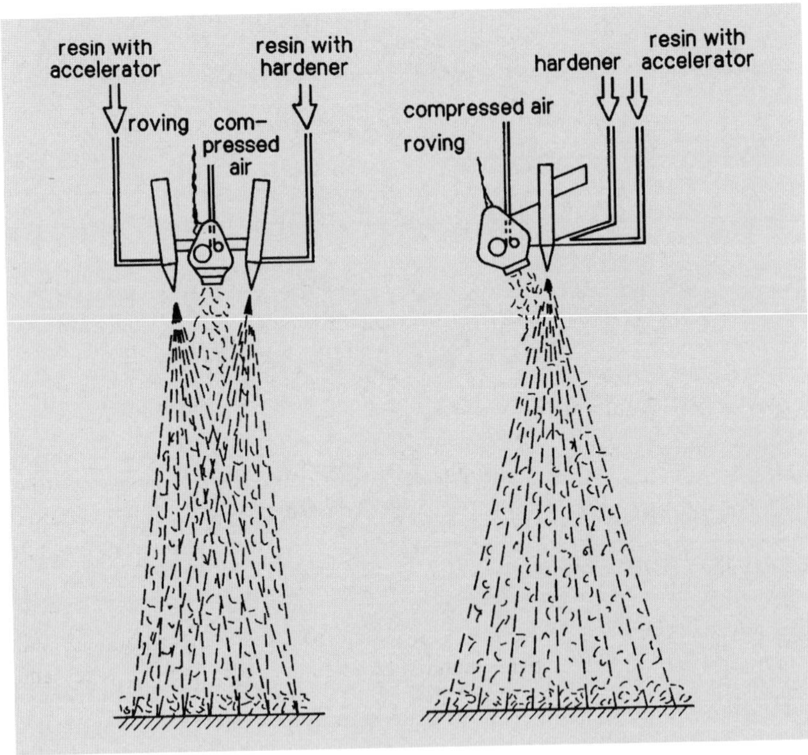

Figure 13-4 Fiber spray-gun molding

application

Because environmentally harmful vapors (e.g., styrene gases) are produced by spraying, it is advisable to use robots working in gastight compartments. Nonetheless, this process is still frequently performed by manual methods. Fiber spray-gun molding is used to manufacture bathtubs, for example.

Winding

process

In the winding process, the fiber strands ("rovings") are impregnated with resin and then wound on a revolving mandrel. The device that guides the filaments—the "payoff eye"—is operated horizontally. The mandrel is thus covered with filaments in the desired manner. This method is represented in Figure 13-5.

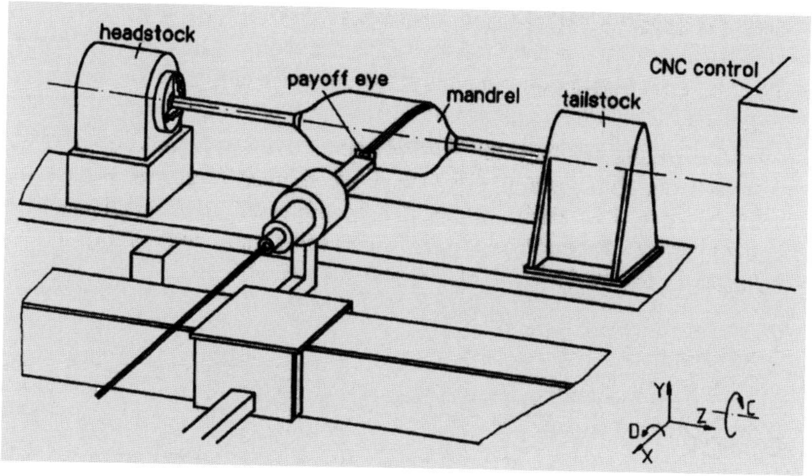

Figure 13-5 Winding machine

The guidance of the rovings and the rotational speed of the mandrel must be precisely regulated. For one thing, the filaments slide off the mandrel if they do not lie in the correct position. Furthermore, the filaments must lie in exactly those directions planned in the design so that they will be able to absorb the forces to be encountered in later use.

filament direction

For complicated parts, it is possible to perform the filament guidance from point to point by hand and store the guidance information in a computer. In the automated production process, this computer then controls a robot, instructing it to repeat the sequence of points.

robots

The advantage of this process is that it can be effectively automated and reproduced. Examples of parts manufactured in this way include tubes and pressure vessels.

application

Compression Molding

Large, relatively flat parts with good mechanical properties can be manufactured by compression molding. Compression molding involves the processing of SMC or GMT molding compounds. SMC (sheet molding compound) consists of a resin preparation with chopped and/or continuous filaments, which later cure to form a thermoset matrix. In GMTs (glass-mat-reinforced thermoplastics), the matrix consists of a thermoplastic material.

process

compression molding compounds

compression molding

Blanks to be used in compression molding are cut from the SMC and GMT semifinished products (webs) and are stacked to form packages. The blank package is laid in the mold of the press, which is then closed and subjected to pressure. This causes the material to flow into all corners of the cavity and fill it. The compression molding method for SMC is represented in Figure 13-6.

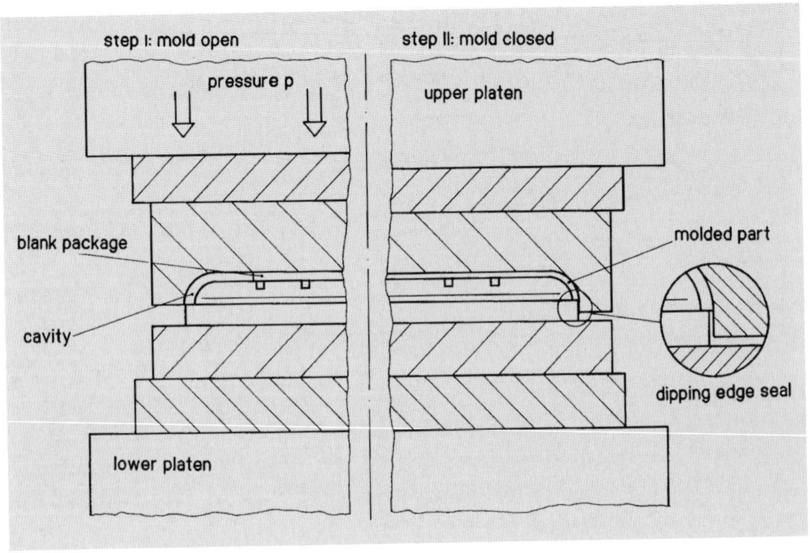

Figure 13-6 Compression molding of SMC

mold

In SMC, the mold is heated and within the material this heating initiates the chemical reaction that cures the part. The GMT plastic still exists in a molten state at the temperature at which it is introduced into the mold. The cooler temperature of the mold causes the plastic to solidify.

part properties

The form of the blank package and its position within the mold are important to the characteristics that the part will later exhibit. Both of these factors influence the flow characteristics of the plastic within the mold, thus affecting the alignment of the fibers as well. This alignment affects the properties of the part, in turn.

application

Compression molding is used to manufacture panels for control cabinets and automobile hoods, for example.

Pultrusion

The pultrusion or string-pull process is available for continuous big-series technical production of continuous filament reinforced profiles. At the moment, the industry exclusively uses thermosetting resin systems, such as polyester or vinylester (the thermosetting fiber-reinforced plastics are currently at research stage). Figure 13-7 shows a standard pultrusion plant. From left to right, the following production steps are shown: impregnation (impregnating bath), consolidation, curing, calibration (mold), and take-off.

Review Questions

No.	*Question*	*Choices*
1	In fiber-reinforced plastics, the plastic that holds the fibers is called the ____________________.	matrix weave mat
2	The modulus of elasticity for steel is 210,000 N/mm^2 (30,450,000 psi). In contrast, the modulus of elasticity for carbon fibers can be as high as ____________ N/mm^2 ____________ psi.	N/mm^2 (psi) 5000 (725,000) 87,000 (12,615,000) 490,000 (71,050,000)
3	The preferred direction of loading for a glass mat is ____________.	 
4	The basic steps in the processing of fibrous composite plastics are: a) application and alignment of the fibers, b) ____________ the fibers, c) ____________ the part, d) ____________ the plastic.	chopping shaping curing impregnation
5	Boats made of fiber-reinforced plastics are often manufactured by the method of ____________________.	winding hand lay-up compression molding
6	The SMC method and the GMT method are both compression molding methods.	
	a) In the SMC method, a ____________ matrix is used.	thermoplastic thermosetting
	b) In the GMT method, a ____________ matrix is used.	thermoplastic thermosetting

Lesson 14

Plastic Foams

Key Questions What are the properties of plastic foams?
How are plastic foams manufactured?

Subject Area From Raw Plastic to Finished Product

Contents

Prerequisite Knowledge Classification of Plastics (Lesson 5)

14.1 Properties of Foams

Fundamentals

gas bubbles

volume fraction

The term "plastic foams" denotes plastics in which gas bubbles are entrapped. The space occupied by the gas bubbles in a foam of this type can amount to 95%, whereas the actual plastic makes up only about 5%. Let's consider the example of a cube with a volume of 1 dm^3 (approx. 64 in^3). If this cube were made of compact polystyrene, it would weigh approximately 1 kg (2.2 lb). A foamed polystyrene cube of the same volume would weigh only 20 g (approx. 3/4 oz).

open cells

If the gas bubbles are connected to one another, the foam is described as an "open-cell foam" (see Figure 14-1).

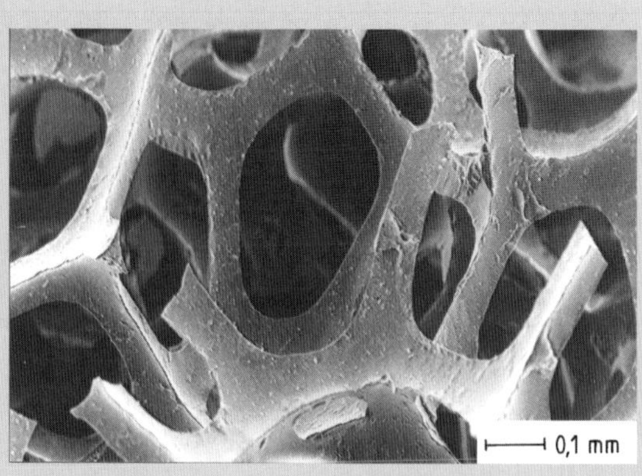

Figure 14-1 Open-cell foam

closed cells

In a "closed-cell foam" each gas bubble exists separately and possesses its own "skin" (see Figure 14-2). Lying between these two extremes are zones of gradual transition, in which the foam exhibits both closed and open cells.

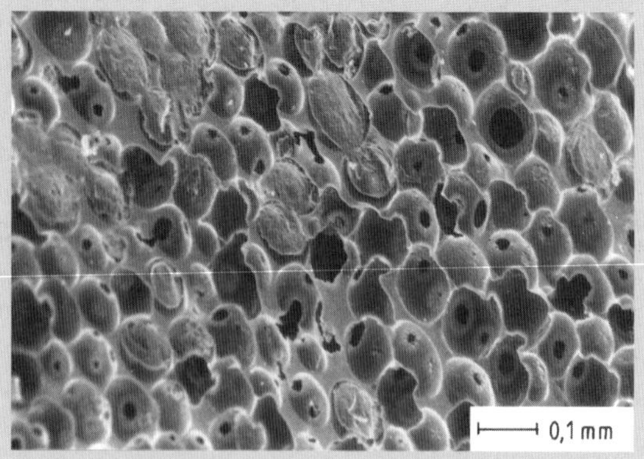

Figure 14-2 Closed-cell foam

The cell distribution can differ from one type of foam to another. This fact is illustrated by the sample comparison between a polyurethane foam and a structural foam in Figure 14-3.

cell distribution

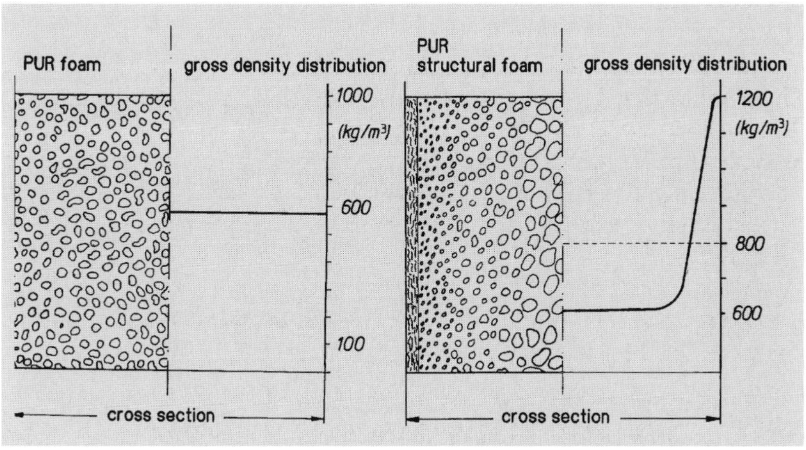

Figure 14-3 Cell distribution in PUR foam and structural foam

PUR foam

structural foam

In the PUR foam the cells are uniformly distributed over the cross section. This results in a similarly uniform density distribution. In contrast, the structural foam displays an uneven cell distribution. Although there are many cells in the middle of the cross section, their number decreases in the direction of the edge. The outermost layer consists almost entirely of compact plastic. This cell distribution is produced by a special foaming technique. As a result of their compact outer skin, the parts manufactured by this technique exhibit great rigidity but are still very light.

Figure 14-4 provides an overview of the densities of the various plastic foams.

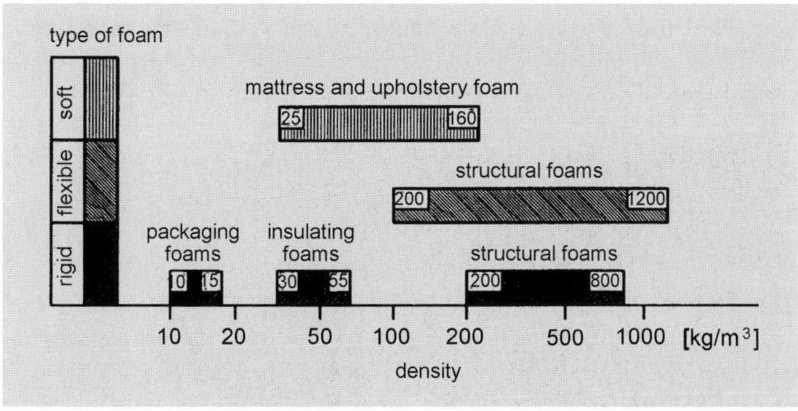

Figure 14-4 Density of foams

Plastics Used in Foaming

Almost all plastics are theoretically suitable for foaming, but only a few them are actually used. Figure 14-5 provides an overview of the plastics an processes involved in foaming.

process	activation	type of reaction	foaming method	cell structure	examples
injection molding	thermal	softening/ cooling	chemical	closed	PVC, PE
extrusion	thermal	softening/ cooling	chemical	open	PVC, PE
multi-component system	mixture	poly-addition	chemical/ physical/ mechanical	open/closed	PUR
multi-component system	thermal	poly-addition	chemical	closed	PA, EP
two-step sintering process	thermal		physical	closed	PS–E

Figure 14-5 Plastic foams

properties

Regardless of the plastic and process used, plastic foams exhibit the followin properties:

— Low density,

— low thermal conductivity,

— favorable mechanical properties at a specific weight,

— simple, yet varied shaping options,

— easy fabrication,

— less material used.

Rigidity of Foams

rigidity

flexible foams

Rigidity is one characteristic of foams. Figure 14-6 provides an overview the various plastics and the rigidity they exhibit after foaming. A "flexib foam" can be deformed easily. When the load is removed, it recovers i original form. Rigid foams can be categorized as ductile-rigid and brittle-rigi When a ductile-rigid foam is subjected to loading, it becomes deformed befo breaking.

foamed plastic	rigidity range
thermosets	
polyurethane (PUR)	ductile-rigid to flexible-elastic
phenol-formaldehyde resin (PF)	brittle-rigid
thermoplastics	
polyethylene (PE)	ductile-rigid to flexible-elastic
polypropylene (PP)	ductile-rigid
polystyrene (PS)	ductile-rigid

Figure 14-6 Rigidity of foams

14.2 Foam Production

Fundamentals

blowing agents
additives

In foam manufacturing, blowing agents and, in many cases, other additives are added to the plastic. These constituents must be mixed very thoroughly to prevent defects and irregularities in the foam. As foaming begins, the mixture must be able to flow freely. Once the bubbles formed by the blowing agent have attained the desired size, they must be fixed. This is accomplished through the hardening of the plastic.

thermosets

thermoplastics

At the beginning of the foaming process for thermosets, the plastic exists as a resin with little or no cross-linking and low inherent viscosity. The fixation of the developing bubbles is caused by the reaction and resulting cross-linking of the plastic. The viscosity of the plastic increases quickly in this reaction. On the other hand, thermoplastics must be melted in order to be foamed. The bubbles become fixed as the plastic cools and solidifies.

foaming mechanisms
mechanical/
physical/chemical

The foaming mechanism that causes the bubbles to develop can be categorized as mechanical, physical, or chemical, according to its respective cause. In a mechanical foaming process, the bubbles are created by using an agitator to stir a gas into the mixture or by using high pressure to force the gas into the plastic melt. In a physical foaming process, heat causes a low-boiling liquid to evaporate, thus forming the bubbles. In a chemical foaming process, the blowing agent reacts under the influence of heat-releasing gases which form the bubbles.

Technical Processes

mixing

Two processes are used to mix the constituents for foaming.

Low-Pressure Process

low-pressure process

One alternative is the low-pressure mixer, in which mixing is performed b mechanical stirring. The advantage is that this process requires only th pressure needed to convey the constituents through the lines. On disadvantage is that only a relatively small quantity can be mixed an conveyed per unit of time. The process is therefore unsuitable for plastics th react very quickly. Another disadvantage is that the mixture will only flow o of the mixing chamber under the force of its own weight. As a result, the on molds that can be used for this foaming process are those into which th material can be poured without additional pressure.

High-Pressure Process

high-pressure process

The other alternative for mixing is the high-pressure mixer. The constituents o the plastic collide with one another under high pressure inside the mixir chamber and are thus mixed. The advantage is that it allows mixing of eve fast-reacting plastics, because the throughput per unit of time is very high. Th mixture does not begin to react until it has been introduced into the mol which occurs quickly. The pressure even makes it possible to use closed mol into which the mixture must be injected. The disadvantage is the high technic expense incurred in developing the required high pressure. Figure 14-7 show a comparison of the two mixers.

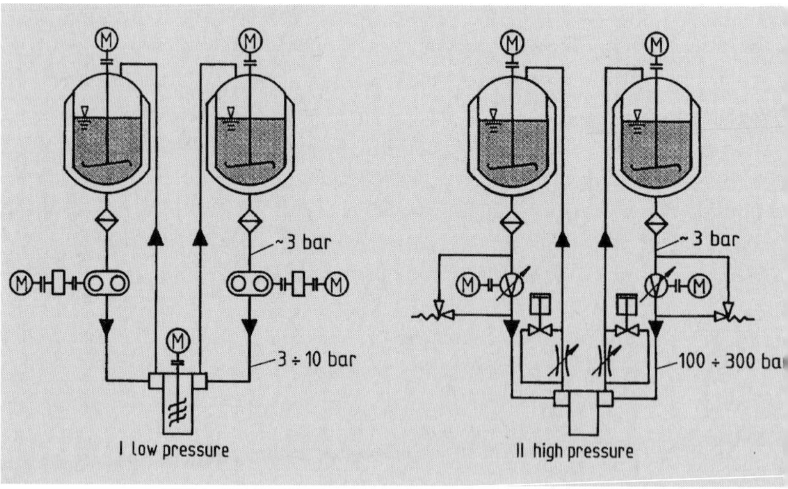

Figure 14-7 Low-pressure and high-pressure mixers (1 bar ≅ 1 atm)

molds

Various types of molds are used to make parts from plastic foams. Fo semifinished products from which upholstery or insulation will be manu factured, the mold takes the form of a continuously operating paper trough open at the top. This paper trough can be seen in Figure 14-8.

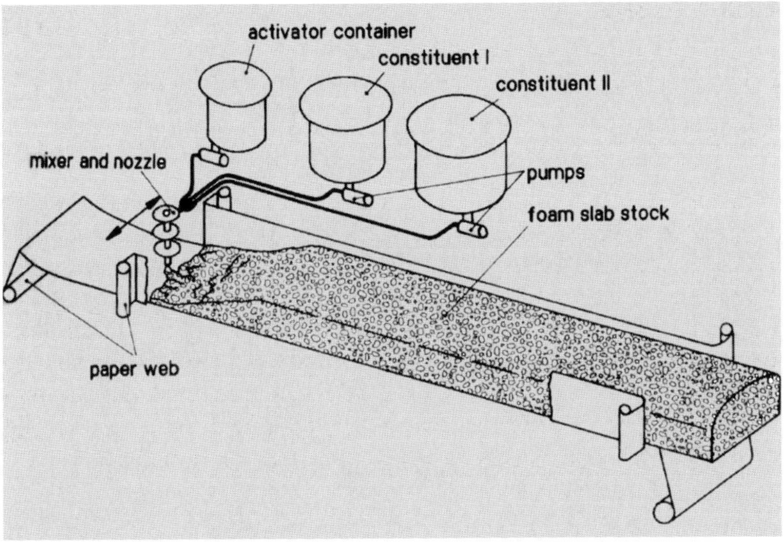

Figure 14-8 Slab stock machine

RIM

Another alternative is a mold similar to an injection mold. The mixture is injected by a high-pressure mixer until one third of the mold has been filled. Then the mixture begins to foam and completely fill the mold. This process is known by the abbreviation RIM, which stands for "reaction injection molding." This name indicates that the development of the molded part occurs through a combination of injection and reaction. Products manufactured by this process include motor vehicle dashboards, for example.

applications

Review Questions

No.	*Question*	*Choices*
1	In plastic foams, ______________ are entrapped within the plastic.	fillers gas bubbles
2	Plastics foams are ______________ compact plastics.	lighter than heavier than exactly as heavy as
3	The gas bubbles of PUR foams are distributed ______________ throughout the plastic.	uniformly nonuniformly
4	There is a considerably ______________ number of cells in the middle of a structural foam than at the edge.	lower higher
5	All plastic foams display the same rigidity. ______________	true false
6	In the manufacture of plastic foams, a distinction is made between mechanical, physical, and chemical ______________.	mixtures foaming methods
7	Even fast-reacting plastics can be processed by the ______________ pressure process.	low high
8	The RIM process is similar to ______________.	extrusion injection molding

Lesson 15

Thermoforming

Key Questions

What steps are involved in the thermoforming process?
Which plastics lend themselves to thermoforming?
What different processes exist?

Subject Area

From Raw Plastic to Finished Product

Contents

Prerequisite Knowledge

Classification of Plastics (Lesson 5)
Deformation Behavior of Plastics (Lesson 6)

15.1 Fundamentals

thermoforming

sequence of the process

The term "thermoforming" denotes the reshaping of plastics under the influence of heat and pressure or vacuum. There are many techniques for performing this process. The use of compressed air and/or vacuum has become the preferred method for reshaping thermoplastics. The general sequence of the process can be described as follows. The plastic is heated to a temperature at which it becomes thermoelastic or rubber-elastic (see Figure 15-1). It is then reshaped and cooled.

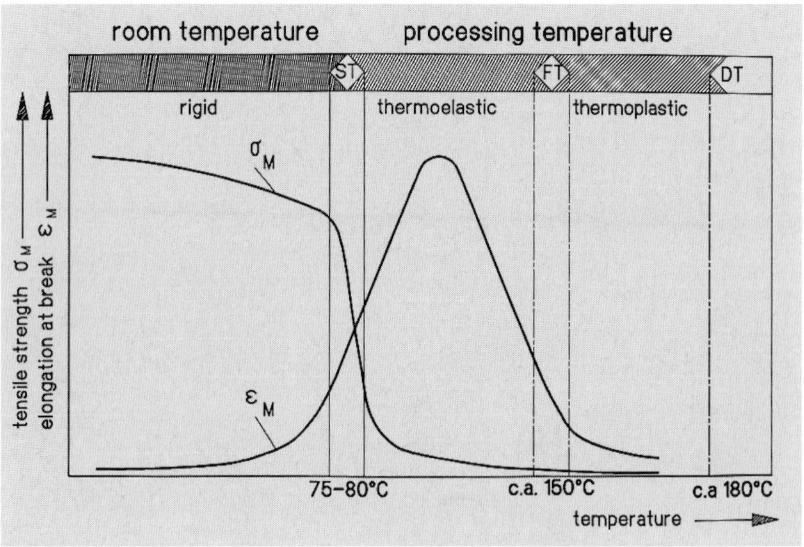

Figure 15-1 Physical state diagram for amorphous thermoplastics

thermoplastics

Because thermoplastics can be converted from the solid state into the thermoelastic range by heating, they are the only type of plastic that can be processed in this manner. In contrast, thermosets do not become rubber-elastic when heated. Therefore, they are not amenable to further processing by thermoforming.

semifinished product

Films and sheets with thicknesses from 0.1 to 12 mm (4 mil–47 mil) are processed most commonly by thermoforming. The material—also known as the "semifinished product"—takes the form of individual sheets or coiled rolls.

15.2 Process Steps

process steps

heating methods

infrared radiation

The process occurs in three steps: heating, forming, and cooling. In the first step, the semifinished product is heated. This can be accomplished by convection, contact, or infrared radiation. The most frequently used method is infrared radiation because it causes energy to directly penetrate the internal regions of the plastic. The plastic is thus heated very quickly and evenly, and its surface is not damaged by overheating.

The second step is the forming of the part. The plastic is stretched in this step. The heated semifinished product is fixed in a clamping device and then pressed into or onto a mold by means of compressed air or drawn into or onto the mold by means of vacuum. A disadvantage of this method is that only one side of the molded part is ever formed with precision, namely, the side that contacts the mold. For this reason, a distinction is made between positive and negative processes, according to whether the inner or outer side of the part is formed with precision. The negative process is represented in Figure 15-2.

forming

types of processes

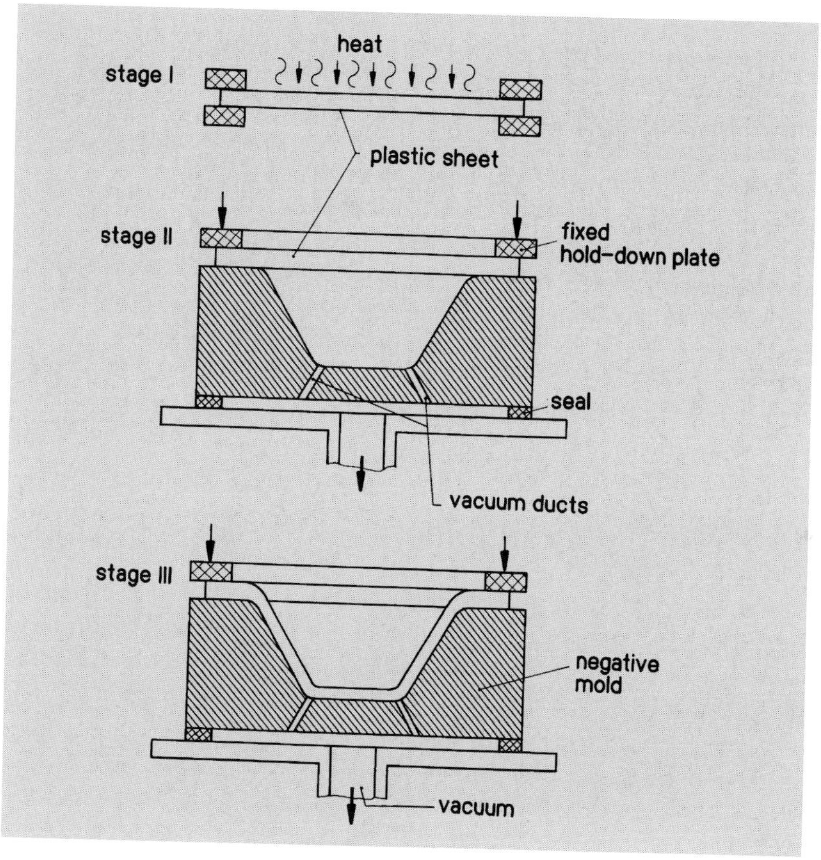

Figure 15-2 Negative molding process

In the negative process, the semifinished product is drawn *into* the mold. In the positive process, the semifinished product is drawn *onto* the mold. During this procedure, the semifinished product is held in a clamp and becomes stretched. This causes the parts to display unequal wall thicknesses. Corner walls are particularly thin.

negative process

In order to minimize this effect, the semifinished product is often prestretched before the actual forming step. In the negative process, this prestretching is performed by a ram. In the positive process, it is accomplished by "inflating" the semifinished product. Figure 15-3 shows the positive process with prestretching.

prestretching

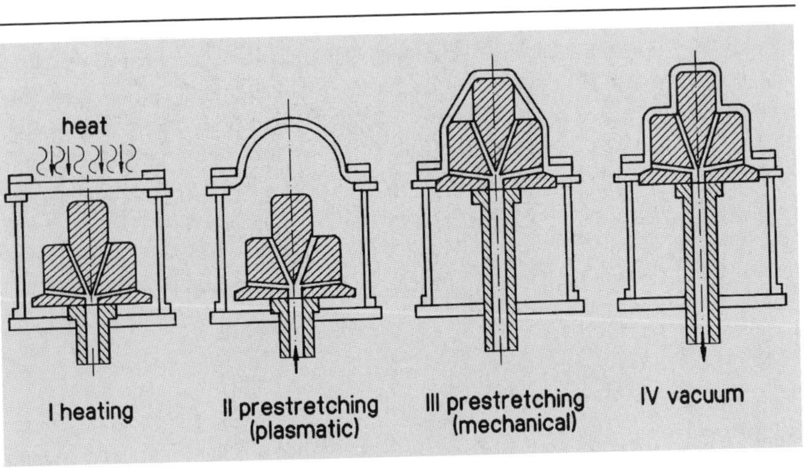

Figure 15-3 Positive thermoforming process with prestretching

cooling

The third step, cooling, begins as soon as the heated semifinished product touches the cooler mold. In order to shorten the cooling time, we can provide the mold with an additional cooling mechanism (for the purpose of series production, for example). Cooling can be further accelerated by cooling the side of the molded part facing away from the mold. This cooling can be accomplished with a blower, for example.

15.3 Technical Equipment

single-station machine

The process steps are technically implemented with either single-station or multistation machines. In a single-station machine, the technical devices are in motion, and the semifinished product maintains the same position all the way from heating until removal from the mold (see Figure 15-4).

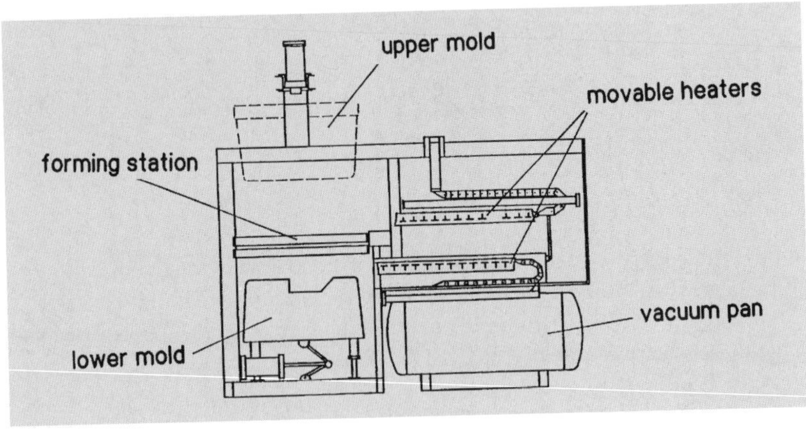

Figure 15-4 Single-station machine

multistation machine

In the multistation machine, the semifinished product moves continuously from one technical station to the next (Figure 15-5).

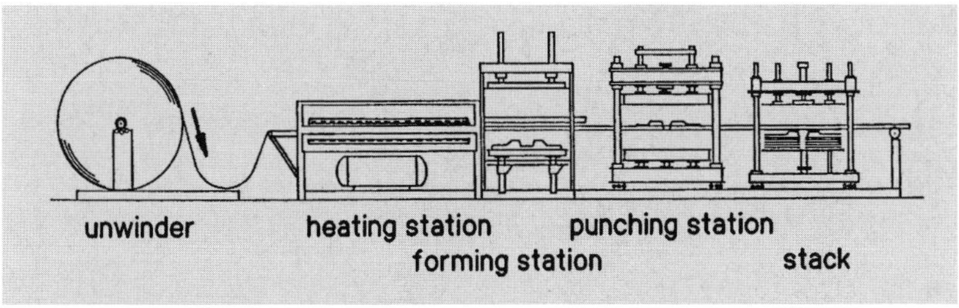

Figure 15-5 Multistation machine

cycle time

The disadvantage of the single-station machine is its long cycle time, which equals the sum of the times required for all individual steps. The cycle time of the multistation machine equals the time required for the longest individual step.

applications

Thermoforming processes are used on a large scale for the manufacture of packaging (e.g., yogurt cups) and for large parts, such as swimming pools or motor vehicle parts.

Review Questions

No.	*Question*	*Choices*
1	In thermoforming, the plastic must be ______________________ before it can be reshaped.	cooled heated melted
2	Only ______________________ can be processed by thermoforming, because they are the only type of plastic to become rubber-elastic when heated.	thermoplastics elastomers thermosets
3	The heating method used most frequently in thermoforming is ______________________.	convection contact infrared radiation
4	In thermoforming, ______________________ of the part is/are formed precisely.	only one side both sides
5	In order to avoid unequal wall thicknesses, the semifinished product is ______________________ before it is formed.	preloaded prestretched clamped
6	The cycle time of multistation machines is ______________________ the cycle time of single-station machines.	shorter than longer than equal to

Lesson 16

Welding Plastics

Key Questions

How does the welding of plastics work?
Which plastics can be welded?
What technical methods are used to weld plastics?

Subject Area

From Raw Plastic to Finished Product

Contents

Prerequisite Knowledge

Classification of Plastics (Lesson 5)
Deformation Behavior of Plastics (Lesson 6)

16.1 Fundamentals

definition

"Welding" plastics denotes the use of heat and pressure to join two parts made of the same or very similar plastics. The surfaces to be joined, also called "mating surfaces," are converted to a thermoplastic (i.e., molten) state for welding. The surfaces are then joined under pressure, and the joint is cooled until its form is stable. The fact that the mating surfaces must be molten indicates that neither elastomers nor thermosets, but only thermoplastics, can be welded.

thermoplastics

16.2 Process Steps

energy supply

welding methods

For the thermoplastic to be fused, energy must be supplied to it. This can be accomplished by five general methods, which are based on different physical processes. The various methods of welding are divided into these five categories (see Figure 16-1).

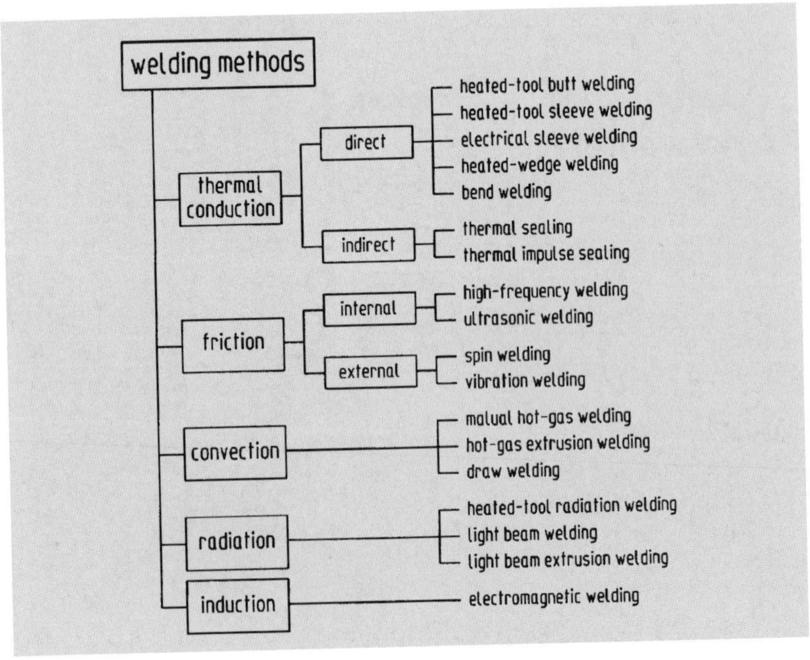

Figure 16-1 Classification of welding methods

pressure

heating time

Aside from the supply of energy to the zone of contact, pressure is also very important. Pressure causes the melt to flow and the two surfaces to become inseparably joined. It is necessary to melt enough plastic to allow the material to mix well. For this reason, the heating time is also very important.

Welding generally consists of five steps:

process steps

Step I: Cleaning the surfaces

Step II: Heating the surfaces

Step III: Applying the pressure

Step IV: Cooling under pressure

Step V: Finishing the welding seam

The ability of two thermoplastic parts to be welded to one another, forming a strong bond, depends on three conditions. (1) They must enter a molten state in approximately the same temperature range so that they both melt at the same time. (2) They must display similar viscosity in the molten state so that they flow easily into one another. (3) The two thermoplastics must be miscible in the molten state. This last requirement implies that welding usually takes place between two parts of the same thermoplastic.

temperature range

viscosity

16.3 Welding Methods

Hot-Tool Welding

A common feature of all methods of hot-tool welding is the use of heating elements to supply heat to the mating surfaces. These tools, usually metal and electrically heated, pass heat to the plastic by thermal conduction. A basic distinction is made between direct and indirect methods of hot-tool welding.

methods

In the direct method, the heat flows directly from the heated tool into the mating surface. In the indirect method, external heat is conveyed to the mating surface through the rest of the part. Because of the poor thermal conductivity of plastics, the indirect method is used only for very thin-walled parts (i.e., films).

direct
indirect

As an example, let's examine the direct method of hot-tool butt welding and the indirect method of thermal-impulse sealing.

Hot-Tool Butt Welding

Hot-tool butt welding is a frequently used method for plastics welding. It is used to join PP and PE tubes, for example, or (as an automated process) to manufacture automobile taillights. The welding process is shown in Figure 16-2.

hot-tool butt welding

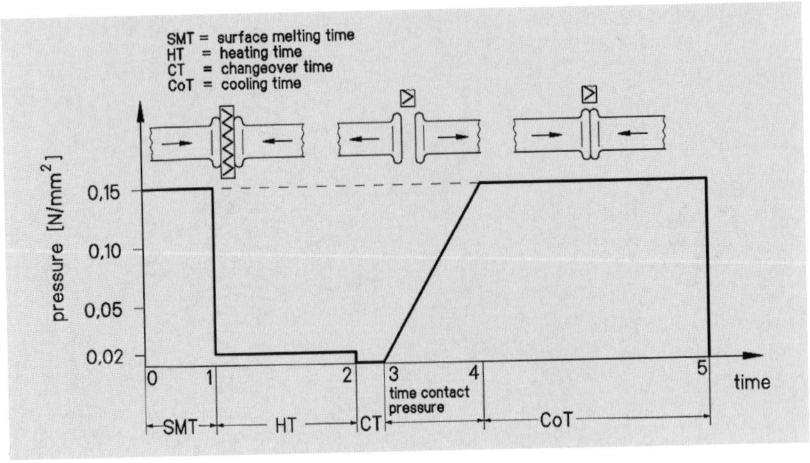

Figure 16-2 Hot-tool butt welding process

surface-melting time

Surface-melting time: The surfaces to be joined are mated by surface melting. A pressure of 0.15 N/mm^2 (22 psi) is applied until a closed bead is visible around the circumference of the mating surface.

heating time

Heating time: The surface-melting pressure is reduced to 0.02 N/mm^2. For the surfaces to be welded, they must be melted with the hot tool by using the reduced contact pressure.

changeover time

Changeover time: The hot tool is withdrawn as quickly as possible.

cooling time

Cooling time: The surfaces to be joined are moved together until they nearly touch. After that, the joining pressure increases, which has to rise quickly from 0 to the final grade. The bulge has to be equal and round. The pressure is maintained until the welding zone is just warm to the touch.

Thermal-Impulse Sealing

thermal-impulse sealing

Thermal-impulse sealing is the most widely used method of indirect hot-tool welding. Because of the poor thermal conductivity of plastics, this method is used only for very thin films. Its greatest area of application is in the packaging industry, where it is used to close pouches, bags, and sacks. The process is shown in Figure 16-3.

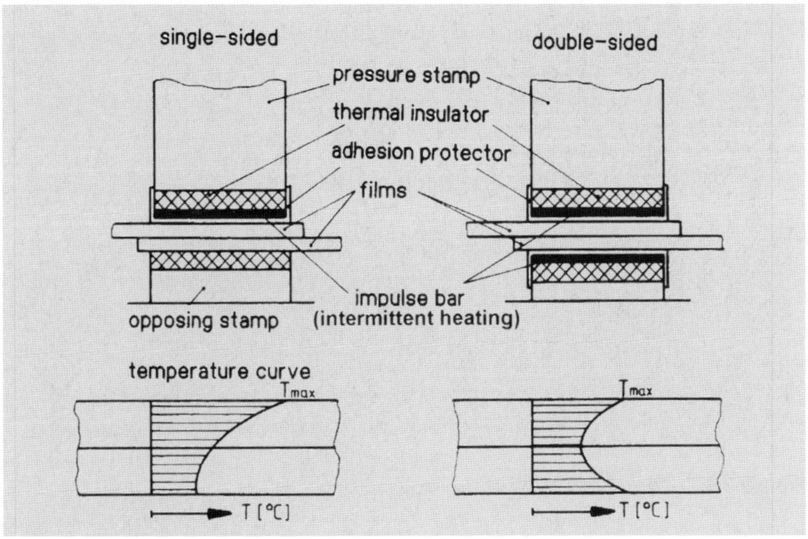

Figure 16-3 Thermal-impulse sealing

thermal conduction

Welding is performed by heating thin metal bars with a brief pulse of high current. These bars, which are provided with an anti-adhesion (release) coating, pass the heat to the films by thermal conduction. The films melt and weld. Both single-sided and double-sided methods of thermal impulse sealing are used. In the single-sided method, the films are heated from only one side by a single metal bar. In the double-sided method, the films are heated from both sides.

distribution of heat

These methods bring about an unfavorable distribution of heat within the parts being welded. It is necessary to attain the melt temperature at the point of contact between the films without allowing the warmer (outer) edge to reach the decomposition temperature of the plastic.

Hot-Gas Welding

hot-gas welding

Hot-gas welding represents another group of welding methods. These methods are usually performed by hand and require a great deal of manual dexterity. The surfaces to be joined are heated by hot gas (e.g., clean compressed air) and welded under pressure, usually with additional material. This method is often used for assemblies and repairs in the construction of apparatus and containers.

manual hot-gas welding

One example of hot-gas welding is manual welding (see Figure 16-4). While the front of the material is heated by the back-and-forth movement of the current, the welding rod must be fed vertically under pressure from above.

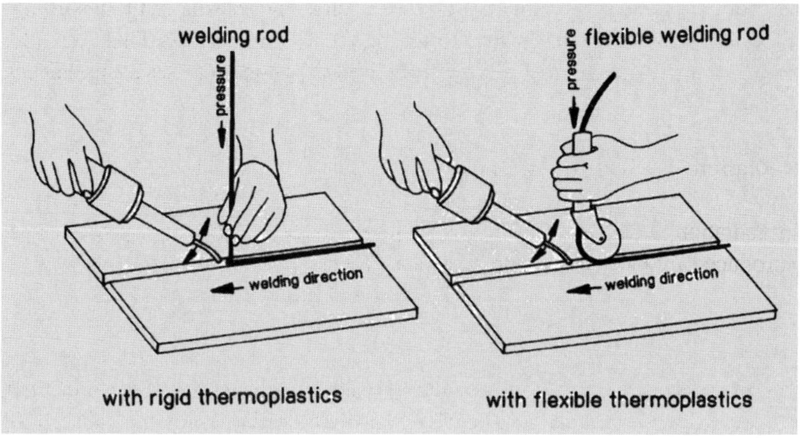

Figure 16-4 Manual welding

Friction Welding

friction welding

Friction welding methods take advantage of frictional heat to melt the plastic. A distinction is made between external and internal friction.

Spin Welding

external friction

In spin welding, parts that are rotationally symmetrical are welded by external friction. While one part rotates, the other remains stationary and is pressed against the rotating part with a certain amount of force. The mating surfaces adapt to one another by melting. Once a sufficiently large bead has developed at the seam, the clamping fixture is released, and the seam cools under pressure. The process is illustrated in Figure 16-5.

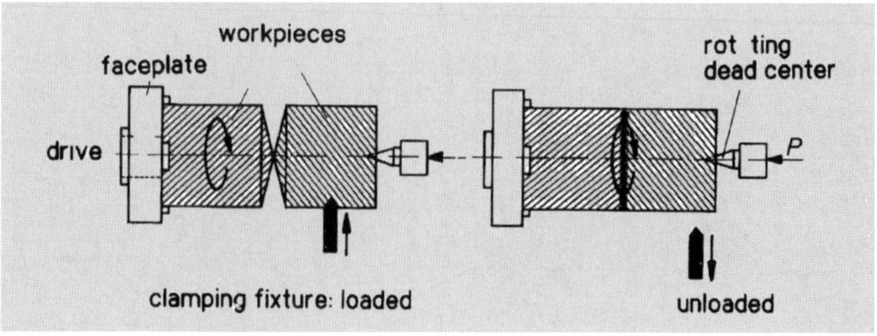

Figure 16-5 Spin welding

Ultrasonic Welding

internal friction

In ultrasonic welding, the material is melted by internal friction. This welding method takes advantage of the mechanical damping property of the plastic. An apparatus is used to generate a high-frequency mechanical vibration. This vibration penetrates the plastic piece and is reflected by the anvil, thus creating a stationary wave. If damping is too strong, the plastic will absorb the vibration, thus preventing it from reaching the mating surface. This method is used in large-scale serial welding of household goods, electrical appliances, and toys.

Radiation Welding Methods

In radiation welding the energy required to fuse the mating surfaces is introduced by rays of heat or light (and laser).

Light-Beam Welding

light-beam welding

The mating surfaces of the plastic are melted by concentrated beams of light. Of course, this cannot be used with transparent plastics, since they absorb too little light.

Induction Welding

magnetic field

In induction welding a supplementary material is placed between the two parts to be joined. This material contains a powder that can be magnetically activated. The powder is stimulated by a high-frequency magnetic field and thus becomes heated. The powder heats the rest of the supplementary material, which in turn heats the mating surfaces. The mating surfaces are then joined under pressure. Even parts with complicated shapes and practically inaccessible surfaces can be welded by this method.

Review Questions

No.	Question	Choices
1	The mating surfaces of the plastic parts become ________________ in the welding process.	molten adhesive
2	Thermosets and elastomers ________________________ be welded.	can cannot
3	Two different plastics can be welded to one another, provided that they exhibit __________________ in the molten state and similar __________________ and __________________.	viscosity melt temperatures miscibility
4	In hot-tool welding, heat is conveyed to the plastic by __________________.	convection radiation thermal conduction
5	The ________________________ welding method is often used to repair containers.	hot-gas hot-tool induction
6	In friction welding, a distinction is made between ________________ and ________________ friction.	internal upper lower external
7	Even ________________________ parts can be welded by the method of induction welding.	large multiple complicated

Lesson 17

Machining Plastics

Key Questions

What properties of plastics affect machining?
What machining principles result from these effects?
What machining methods and tools are used?

Subject Area

From Raw Plastic to Finished Product

Contents

Prerequisite Knowledge

Classification of Plastics (Lesson 5)
Deformation Behavior of Plastics (Lesson 6)

17.1 Fundamentals

methods

The methods for machining plastics include sawing, milling, turning, drilling, grinding, and polishing.

properties of plastics

The experience gained through the use of these methods in machining metals cannot be directly transferred to the machining of plastics, however. Plastics display properties different from those of metals.

1. Plastics do not conduct heat as effectively as metals. Therefore, the heat resulting from friction during machining does not dissipate readily from the material. The site of the cut must be cooled in an especially effective manner so that the plastic does not melt, let alone decompose.

2. Plastics exhibit a high degree of thermal expansion. When plastics are cut, the saw blade may become stuck. When plastics are drilled, undesirable dimensions may be obtained. After cooling, the bore may turn out to be 0.05–0.1 mm (2–4 mils) smaller than the selected drill bit.

3. Plastics display an especially high degree of notch and crack sensitivity. Machined cuts must be smooth to maintain the mechanical load-bearing capacity of the plastic.

4. Plastics generally exhibit less strength than metals. Machining therefore requires lower forces.

machining principles

These properties give rise to certain principles that should be followed when machining plastics:

1. Thermoplastics should not be heated above 60°C (140°F) during machining. Thermosets should not be heated above 150°C (300°F) during machining.

2. Heating can be controlled through the cutting speed, feed, and tool geometry. It is also possible to use cooling media to cool the site of the cut.

3. Smooth-running machines should be used to create smooth cuts.

17.2 Technical Processes

Sawing

circular saws

High-speed steel or carbide-tipped blades are used in circular saws. The blades must be hollow-ground, with relatively small pitch (see Figure 17-1).

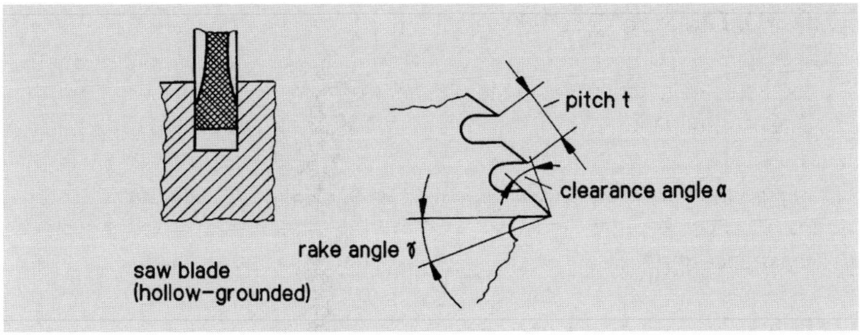

Figure 17-1 Design of saw blade teeth

On band saws, the teeth have a slight set, in order to prevent the tooth spaces from becoming smeared with plastic. Figure 17-2 gives several standard values for sawing plastics.

band saws

plastics	tool	α [°]	γ [°]	t [°]
thermoplastics	HSS (hight-speed steel) carbide	30–40 10–15	5–8 0–5	2–8 2–8
thermosets	HSS carbide	30–40 10–15	5–8 3–8	4–8 8–18

Figure 17-2 Standard values for sawing plastics

Milling

Mill cutters for plastics have fewer flutes than mill cutters for metals but more flutes than mill cutters for wood. They consist of high-speed steel or carbide, or can be carbide-tipped. They should be operated at the highest possible cutting speed and a relatively low feed. The harder the material, the smaller the rake angle should be (Figure 17-3).

mill cutters

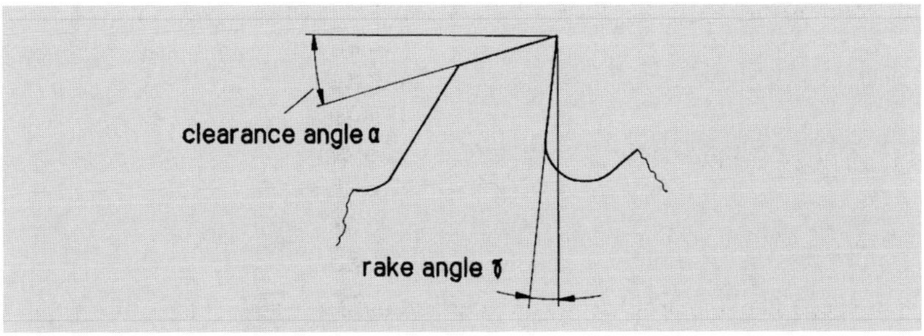

Figure 17-3 Tool angles for mill cutters

For softer materials, a lower number of flutes and greater feed should be selected. Figure 17-4 gives a few standard values for milling plastics.

plastics	tool	α [°]	γ [°]
thermoplastics	HSS (high-speed steel)	2–15	up to 15
thermosets	HSS carbide	up to 15 up to 10	15–25 5–15

Figure 17-4 Standard values for milling plastics

Drilling

drill

Twist drills for metals can also be used for plastics. A steep helix angle allows the bit to remove the chip more effectively (Figure 17-5).

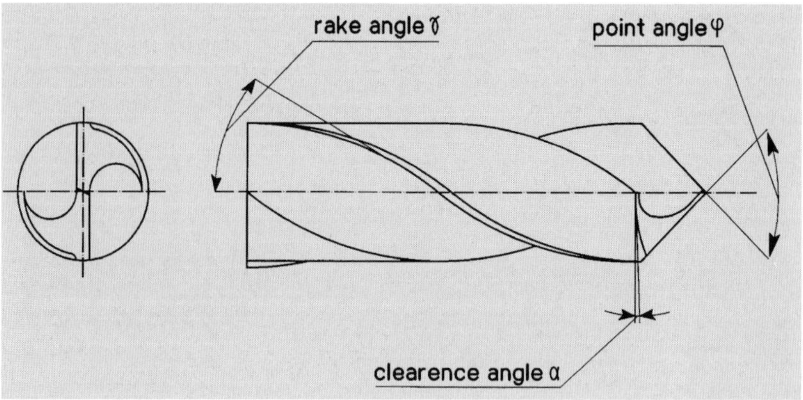

Figure 17-5 Tool angles for twist drills

frictional heat

Because the frictional heat produced by drilling causes plastics to expand significantly, the boreholes turn out to be 0.05 to 0.1 mm (2–4 mils) smaller than the diameter of the drill bit. In practice, a slightly larger bit must be used to produce the desired dimension.

heat removal

Materials that smear easily, such as PE and PP, are machined at a higher feed and lower cutting speed to remove the heat along with the chip. For bore diameters of 10 to 150 mm (0.4–6.0 in), drilling is done with a hollow drill set with a diamond. Several standard values for drilling plastics are given in Figure 17-6.

plastics	tool	α [°]	γ [°]	φ [°]
thermoplastics	HSS (high-speed steel)	3–12	3–5	60–110
thermosets	HSS carbide	6–8 6–8	6–10 6–10	100–120 100–120

Figure 17-6 Standard values for drilling plastics

Turning

turning

The lathe should operate at high speeds and be equipped with a liquid cooling system. The turning tools may be made of high-speed steel depending on the type of plastic. The angles of turning tools for plastics are given in Figure 17-7. Turning tools with carbide edges are used for thermosets and plastics with glass fiber fillers. Several standard values for turning plastics are given in Figure 17-8.

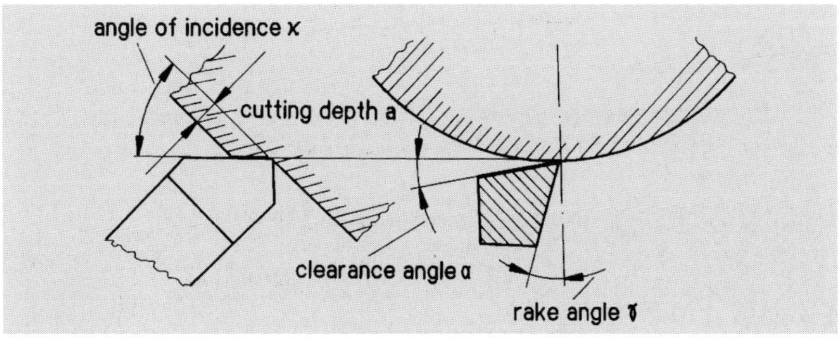

Figure 17-7 Tool angles for turning tools

plastics	tool	α [°]	γ [°]	κ [°]	a
thermoplastics	HSS (high-speed steel)	5–15	up to 10	15–10	up to 6
thermosets	HSS carbide	5–10	15–25 10–15	45–60 45–60	up to 5 up to 5

Figure 17-8 Standard values for turning plastics

Grinding and Polishing

grinding

Grinding is performed with conventional abrasive papers or belts. The grinding speed of the belts should be approximately 10 m/s (2000 ft/min).

polishing

Felt wheels or buffing wheels with polishing compounds are used to polish plastics. To avoid melting the surface of thermoplastics, the polishing process is interrupted frequently.

Review Questions

No.	*Question*	*Choices*
1	As a plastic is being sawn, it may melt at the site of the cut, because it conducts heat ____________ effectively than metal.	more less
2	The high degree of ____________ of plastics can cause the saw blade to become stuck in the plastic while sawing.	thermal conductivity thermal expansion viscosity
3	After the plastic has cooled, a hole drilled in the plastic will be ____________ the diameter of the bit.	larger than smaller than exactly as large as
4	The same drill bits ____________ be used for both plastics and metals.	can cannot
5	Milling should be performed at the ____________ possible cutting speed.	highest lowest
6	The lathe should be equipped with a(n) ____________ cooling system.	liquid air

Figure 18-5 shows several possible forms of adhesive joints.

types of joints

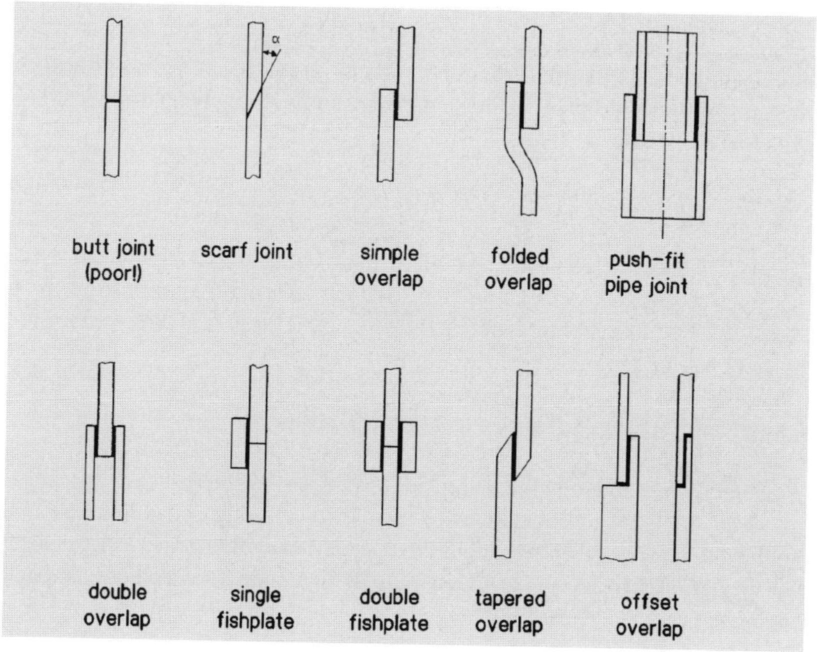

Figure 18-5 Forms of adhesive joints

18.2 Classification of Adhesives

Physically Setting Adhesives

Solvent-Based Adhesives and Dispersion Adhesives

solvents

solvent effects

In order to achieve good wetting of the bonding surfaces, adhesives are often dissolved in an organic solvent or dispersed (i.e., very finely distributed) in water. For the adhesive to become strong and securely cement the bonding surfaces, the solvent must be removed from the adhesive. The solvent either evaporates or is absorbed by the bonding surface. However, it is necessary to determine whether the solvent has a negative effect on the plastic. For example, it is possible that the solvent will release internal stresses in the plastic that may create stress cracks in the part.

contact adhesives

A contact adhesive is one example of a solvent-based adhesive. After being wetted with the contact adhesive, the parts being joined must lie exposed until the solvent has evaporated from the adhesive. The surfaces are not pressed together and bonded until the adhesive feels dry. At this point, it is no longer possible to correct bonding errors.

Solvent Cementing

solvent cementing

One special type of adhesive bonding uses only a solvent that superficially dissolves the plastic. The solvent is applied to the surfaces, diffuses into them, and solubilizes the plastic. When the parts to be joined are pressed together, their respective molecules become entangled, and a strong bond develops. The process can be seen in Figure 18-2.

Hot Melt Adhesives

hot-melt adhesives

Hot-melt adhesives are applied to the bonding surfaces as a plasticized mass. The surfaces are then pressed together, and the adhesive becomes strong as it cools. Because this cooling requires little time, this process is preferred for use in mass production.

Chemically Setting Adhesives (reaction adhesives)

reaction adhesives

As their name suggests, reaction adhesives set by means of a chemical reaction. In this reaction, which can involve addition polymerization, polyaddition, or polycondensation, cross-linked macromolecules (thermosets) are created. The reaction can be started by hardeners, accelerators, or heat, depending on the system.

curing

The different components of the adhesive (in a system of two or more components) cannot be mixed until shortly before processing, because the reaction and curing of the mixed adhesive proceed quickly. The adhesive can no longer be processed after curing.

18.3 Cementing Procedure

quality

The cementing procedure has a decisive influence on the quality of the adhesive bond. As indicated, cementing is performed in the following steps:

Creation of Appropriate Bonding Surfaces

bonding surfaces

The most important requirement in adhesive bonding is that the shapes of the parts being joined and of the joint are suitable for bonding. This will determine the type of stress exerted on the joint by forces encountered in use. As mentioned, such forces should ideally exert no peeling stress on the joint.

Cleaning and Degreasing the Bonding Surfaces

cleaning

It is also important that the adhesive bond be not impaired by impurities (dust, grease, etc).. Depending on the type of impurity involved, cleaning may be performed with continuous baths containing organic solvents or alkaline cleansers, ultrasound baths, or vapor-degreasing baths.

Pretreatment of the Bonding Surfaces

The surfaces are pretreated to further enhance their properties for adhesive bonding. For plastics that are easily cemented, this can be accomplished by mechanical roughening (grinding, sandblasting) or chemical treatment (corrosion). Surfaces of plastics that are difficult to cement are activated by oxygen treatment (flame treatment, corona treatment) or chemical oxidation (corrosion).

pretreatment

Application of the Adhesive

When applying the adhesive, strive for a uniform wetting of the bonding surfaces and a consistent layer thickness.

adhesive application

Waiting Until the Adhesive Is Ready for Bonding

The amount of time before the adhesive is ready for bonding varies widely among adhesives. This waiting time must always be maintained; otherwise the bonding process will be impaired, and the resulting adhesive bond will be markedly weaker or fail entirely.

waiting time

Joining and Clamping the Parts to Be Cemented

After the cemented parts have been joined together, pressure is applied. This pressure displaces the air from between the bonding surfaces and thus determines the thickness of the adhesive film. For adhesives that exhibit a longer curing time, slippage may be prevented by clamping the cemented parts after they have been pressed together.

joining

Curing the Adhesive

The various adhesives exhibit curing times of different lengths. In each case, this curing time represents the length of time that must pass before the adhesive joint can be subjected to loading.

curing

Removal of the Clamp from the Cemented Parts

After the adhesive has cured, the clamp can be removed from the cemented parts. However, although the adhesive has cured thoroughly enough to prevent the parts from slipping, it is often necessary to wait somewhat longer before the adhesive joint is fully capable of withstanding loads.

clamping

When performed correctly, the adhesive bonding of plastics is a valuable method for creating inseparable joints.

Review Questions

No.	*Question*	*Choices*
1	In contrast to welding techniques, adhesive bonding techniques ________________ be applied to elastomers and thermosets.	can cannot
2	Adhesion bonding is based on ________________.	cohesion adhesion cohesion and adhesion
3	In order for the adhesive bond to display great strength, the bonding surfaces must be especially ________________.	large smooth clean
4	The forces that act on the adhesive joint should not have a ________________ effect.	tensile compressive peeling
5	Reaction adhesives become ________________ after curing.	thermosets elastomers thermoplastics
6	Reaction adhesives are not mixed until shortly before processing, because they react ________________ and then can no longer be processed.	quickly slowly

Lesson 19

Plastics Products and Plastics Waste

Key Questions

What types of plastics exist?
What is the volume of plastics produced?
What products are manufactured from plastics?
How long are these products used?
What special considerations affect the handling of plastics wastes by the waste disposal industry?

Subject Area

Ecological Aspects of Plastics

Contents

Prerequisite Knowledge

Classification of Plastics (Lesson 5)

19.1 Discussion of Plastics Wastes

plastics waste

For several years, criticism has been increasingly focused on the topic of plastics wastes. The problems of plastics wastes are especially associated with the following four areas:

problem of volume

1. Plastics wastes occupy a great volume in relation to their weight. They are difficult to compress and thus require quite a large amount of space in disposal sites (landfills).

nondegradability

2. In general, plastics wastes do not degrade readily and therefore do not enter into the biological cycle of decomposition and regeneration.

hazardous substances

3. Some plastics wastes contain substances that create problems when burned in trash incineration plants. Examples are chlorine from PVC, nitrogen from PUR and PA, fluorine from PTFE, sulfur from synthetic rubber, and heavy-metal additives from many plastics.

recyclability

4. Plastics wastes usually cannot be directly recycled because they often occur in a contaminated and mixed state. At present, this often leaves waste disposal businesses no choice but to accept the above-mentioned difficulties in incinerating or disposing of these wastes in landfills.

reduction and utilization

All of these problems would recede into the background if it were possible to reduce plastics wastes and utilize them more successfully. If plastics could be exploited as a valuable material after their initial use, rather than remaining unused in a disposal site or yielding only its energy content through incineration, that would be a further advantage.

The quantity and composition of plastics wastes are examined in greater detail in this lesson to familiarize the reader with possibilities for reducing and utilizing these wastes.

19.2 Plastics in Production and Processing

plastics production

Plastics production has grown steadily since its inception. In the last 30 years, it has increased by close to a factor of 10. North America, Europe and Asia are the major producers of plastics, each region consuming about an equal share of plastic products. Southeast Asia was the region with the highest growth rate in the 90ies. Figure 19-1 shows the world production of plastics according to type in 1998.

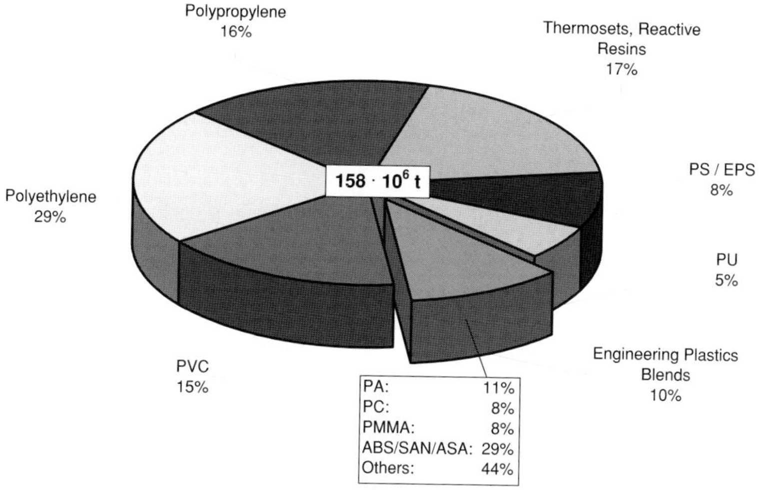

Figure 19-1 World production of plastics by type (1998)

Figure 19-2 shows the distribution of the manufactured plastic products among various areas of application. Packaging and building/construction are the most important sectors of plastics application.

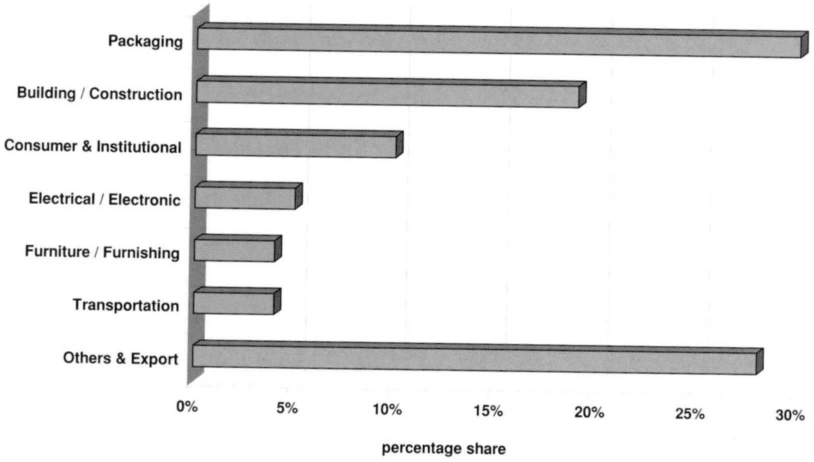

Figure 19-2 Plastic products according to areas of application in the US (1998)

19.3 Plastics Products and Their Life Expectancies

life expectancy

packaging

The life expectancy of plastics products is generally underestimated. The public continues to associate plastics with disposable articles. This can surely be attributed to the use of plastics in nonreturnable packaging. But if we take a closer look at the areas in which plastics products are used, as introduced in the preceding section, we find that packaging accounts for less than a quarter of this use. The predominant applications are those in which plastics are processed to form strong, long-lasting products. This impression is also confirmed by product life studies with results shown in Figure 19-3.

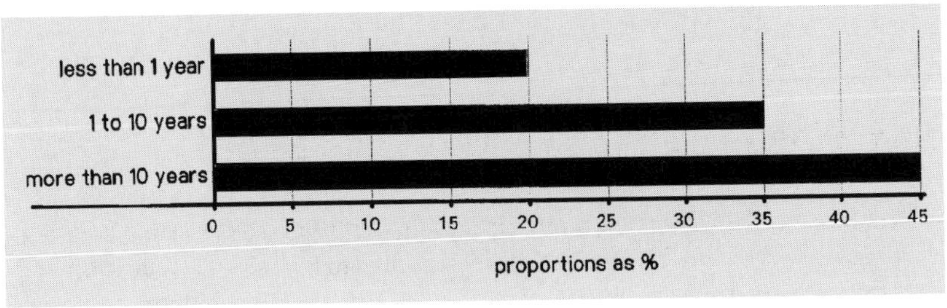

Figure 19-3 Life time of plastics products

period of use

Here we see that approximately 20% of plastic products are discarded within one year, and 35% of all plastic products are used for 1 to 10 years. Another 45% of plastic products do not occur as waste until more than 10 years have passed.

CD

The CD is one example of a long-lasting product. Even the case for the CD will generally be kept for a long time before being discarded as trash. On the other hand, the protective packaging film enclosing the CD case becomes recyclable plastics immediately after the purchase is made. Like most packaging materials, it is used only briefly which, however, can be recycled.

municipal waste

The composition of the plastics in municipal waste in Western Europe is represented in Figure 19-4. We can see that PE, PP, PS, and PVC are found in the greatest amounts in household trash. The polyolefins PE and PP are the most prevalent of these, accounting for 65%. In the U.S. another major component of household trash is PET.

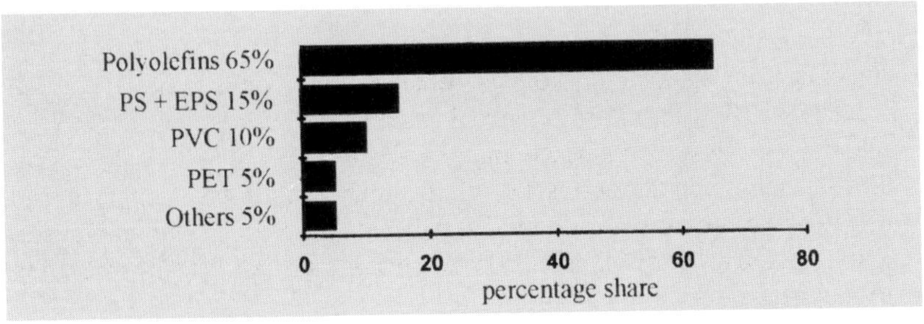

Figure 19-4 Composition of plastics in municipal solid waste by type in Western Europe (1989)

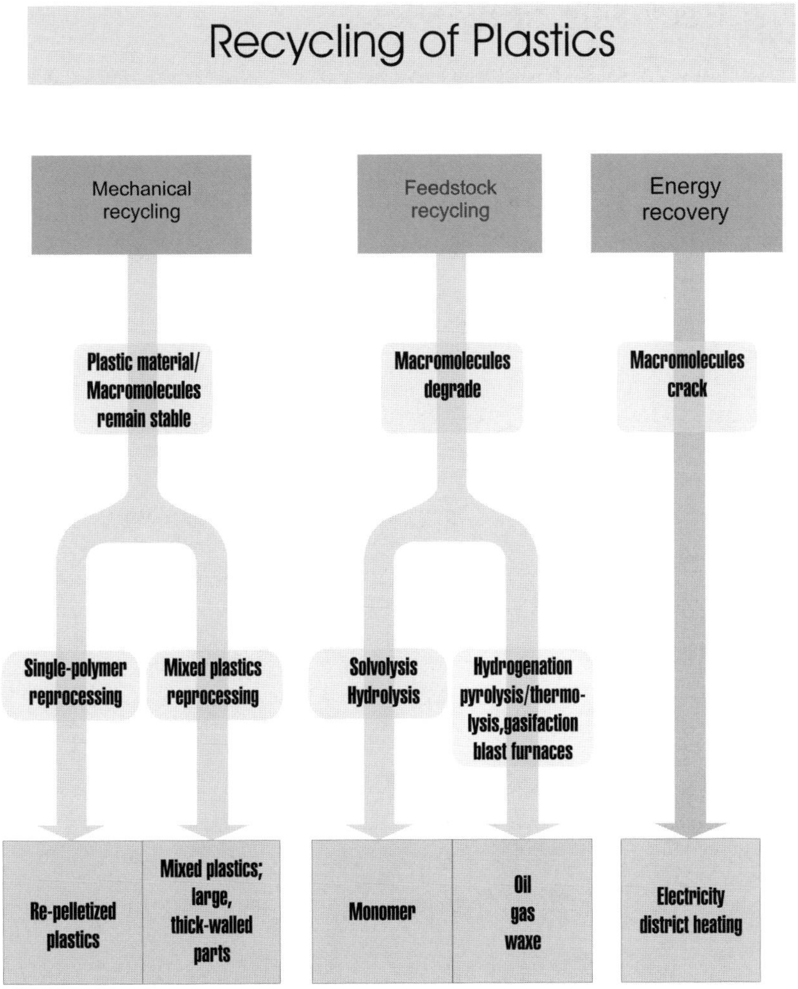

Figure 20-1 Recycling cycles

melting

During the melting process of plastics mixtures, certain plastics degrade due to the temperature applied, whilst other plastics have not reached their melting point, yet. Figure 20-3 shows examples of various melting rsp. softening points of PVC, PA and PC.

melting temperature range

The melting temperature range of PVC is 120–190°C, PA 235–275°C. The melting temperature range for PC, which is used for CDs for example, is 270–320°C. Thus, there is no common melting temperature for a mixture of various plastic material because, at 250°C for example, PVC has already melted, PC has not begun to melt whereas for PA, it is the ideal melting temperature.

impurities

Impurities of waste should be avoided or removed. Otherwise, they will be incorporated into the melt along with the plastic and thus reduce the quality of products. For example, the impurities in yoghurt cups, i.e. the remaining yoghurt, often accounts for a greater proportion of impurity of the overall weight than the cup itself, which weighs only about 6 g (0.2 oz.). When collecting plastics waste, the weight of the impurities is often greater than the actual plastic material. Thus, the plastic has to be cleaned first.

Figure 20-2 Mechanical recycling

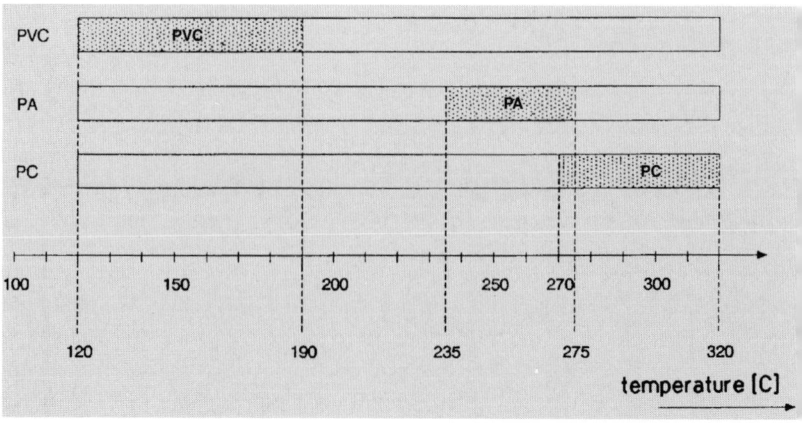

Figure 20-3 Melting (softening) temperature ranges of various plastics

The best results for recycling of thermoplastics can be achieved when all waste is homogeneous, separated by kind of plastic, type of plastic, additives and fillers. Furthermore, plastic should be clean in order to produce better high-quality plastics.

homogeneous plastic

Figure 20-2 shows various existing technologies for reprocessing of plastics waste. They all follow the same principle: size reduction, washing, sorting, drying, and re-pelletizing.

The industry uses different types of plants. Their technologies vary in selection and composition methods. Each type of plant is normally constructed for recycling of certain type of waste (e.g. homogeneous industry waste, used agricultural foils, household plastics, etc.)

Examples of material recycling of technical parts after use are: recycled PET deposit/refund bottles, bottle crates, dye cartridges of the textile industry, control boxes of heating systems, window profiles, tail and indicator lights, and bumpers from cars.

examples

Recycling of Thermosets

Thermosets cannot be melted. However, this contradicts the fact that material can be recycled immediately by remelting. The used material consists of resin, hardeners, fillers, and reinforcing particles. Fillers and reinforcing particles represent the majority of components, of up to 80% of the weight of a material. This fact is used for the recycling of particles to reuse material. The particles can be broken up (ground) and used as fillers in or with new thermosetting material.

fillers, reinforcements

20.3 Feedstock Recycling

Figure 20-4 shows feedstock recycling, using degradative extrusion. Plastics waste is reduced in size and processed to gases, waxes, and oils by using pressure and high temperature in the extruder. These gases, waxes, and oils can be put at the chemical industry's disposal.

degradation extrusion

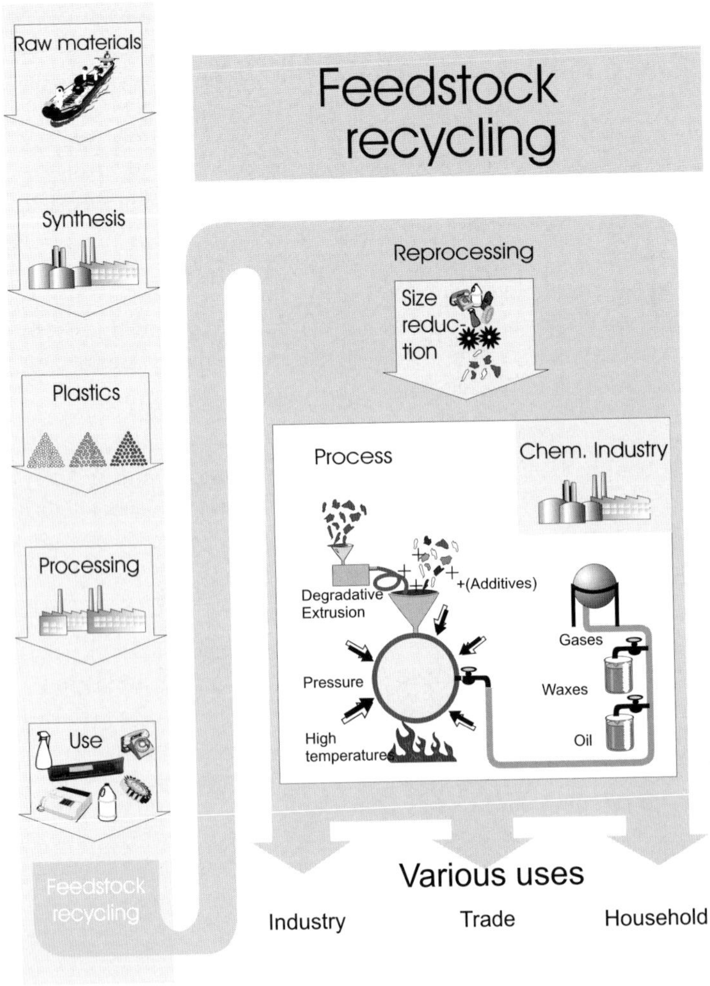

Figure 20-4 Feedstock recycling using degradative extrusion

20.4 Energy Recovery

For parts of plastics waste for the next few years, there is no other form of recycling but energy recovery (Figure 20-5) available because the costs to recycle very impure plastics waste are too high. Thus, no other recycling method is advisable, neither ecologically nor economically. Energy recovery generally means burning. The goal to burn plastics is to use the high energy of plastics, to reduce waste, to inertialize burning remains and to avoid uncontrolled dismissal of harmful substances. Energy recovery is very useful for heterogeneous waste (for example, household waste or car shredding waste) in which plastics represent part of the waste.

energy of plastics

How Can We Recycle a CD?

CD

The CD itself is a composite material consisting of three layers: The layer holding the music information is made of clear PC, the reflecting layer is made of aluminum, and a lacquer layer protects the CD.

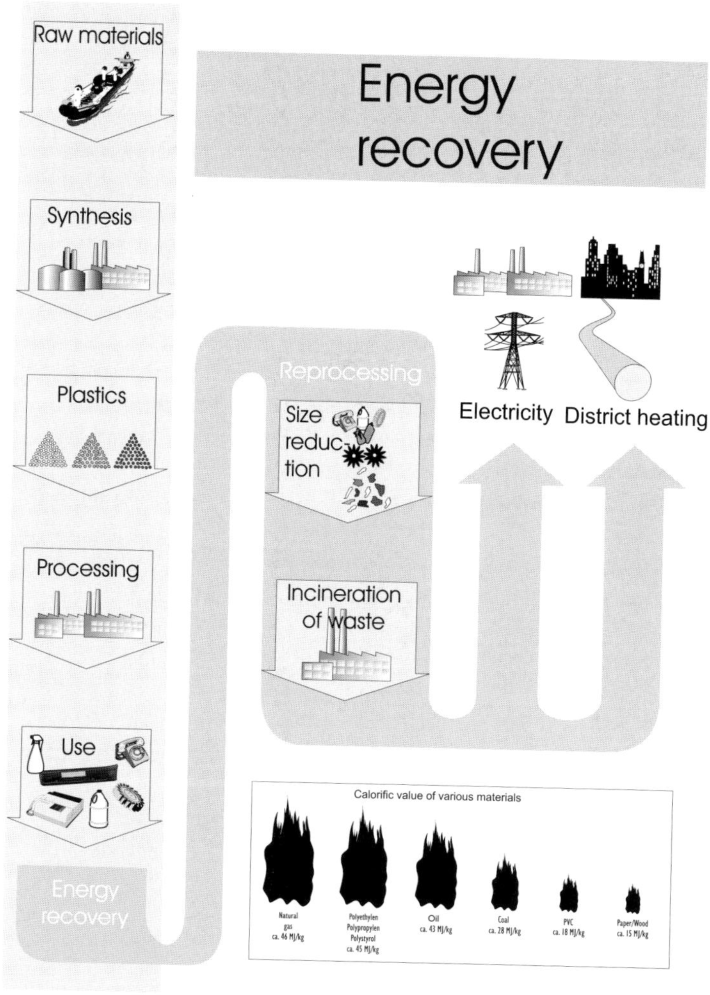

Figure 20-5 Energy recovery of plastics

CD recycling

Bayer AG runs Europe's first CD recycling factory, which allows recycling of polycarbonate on an industrial scale by applying feedstock recycling. Generally, all types of CDs, CD-ROMs as well as other data carriers made of polycarbonate can be recycled this way.

Bayer developed a method to remove the coating from polycarbonate, which does not leave any remains and which is environment-friendly. The pure PC grain, which is left after CD recycling, is being processed to high-quality products. In 1995, about 1,500 t CD waste could be recycled and reused as a high-quality recycling products. This equals roughly the amount of a 100 million CDs.

CD cases

It is different for the three-part CD cases: bottom and lid consist of clear PS. The part that holds the CD, is made of colored PS. The list of song titles contained on the CD, is made of paper. The paper is not glued to the plastic and can be removed. If the parts of the case are sorted by color, new, clear parts can be produced from old clear parts, which have been subjected to size reduction and remelting.

Review Questions

1. Recycling of plastics is ____________________	possible not possible
2. The three possible recycling methods are: mechanical recycling, ____________________ , and energy recovery.	deposit feedstock recycling biodegradability
3. Mechanical recycling contributes to reduction of used ____________________	raw materials energy and raw materials
4. Thermoplastics ____________________ by melting.	can be recycled cannot be recycled
5. Thermoset waste can be ____________________ and reused for the production process.	grained plasticized burnt
6. Plastics waste can be recycled better if used____________________	pure impure
7. Mechanical recycling can be applied to reprocess production waste and to produce high-quality products if plastics waste is pure and____________________	mixed unmixed
8. Unmixed recycling materials fulfill ____________________ quality requirements.	high low
9. During feedstock recycling, gases, ____________________, and oils are being produced.	granules waxes

Appendix I

Glossary of Plastics Technology

ABS acrylonitrile/butadiene/styrene (an amorphous copolymer)

addition polymerization (chain-reaction polymerization) chemical reaction in which polymers are formed from monomers by breaking the double bond (C=C); can be initiated by free radicals, ions and certain organometallic catalysts

aggregate state for plastics, either solid or liquid (plastics decompose before they attain the gaseous state)

amorphous without (regular) form, glassy, noncrystalline, a condition of great disorder or absence of structure

anisotropy condition in which properties depend on direction (i.e., properties are different in different directions)

caloric value quantity of heat resulting from the combustion of 1 kg of solid or liquid fuel or 1 m^3 of gaseous fuel

catalysis acceleration of a chemical reaction by catalysts

cavity specially shaped open space in a mold, which is filled with the material

cellulose most commonly occurring carbohydrate; cotton, jute, flax, and hemp are almost pure cellulose

CFRP carbon-fiber-reinforced plastic (CFRP)–a composite material of carbon fibers and a polymer matrix

chemical bond cohesive forces between atoms in molecules, exerted by ions or pairs of electrons

cross-linking linking together of polymer molecules (macromolecules) by primary valences, resulting, in most cases, in a three-dimensional network; cross-linking of certain plastics can be carried out chemically by the addition of appropriate bridge-building monomers

CRP carbon-reinforced plastic; see CFRP

crystal a solid body with periodically arranged "building units" (atoms or molecules) bounded by flat surfaces, and representing the most orderly possible condition

crystalline consisting of numerous tiny crystals that are not completely formed

crystalline melting temperature (CMT)	crystalline regions of a thermoplastic melt at this temperature
decomposition temperature	the temperature (T_d) at which a material is destroyed by chemical decomposition
delamination	detachment of the fibers from the matrix, or a crack in the matrix running parallel to a laminate layer
dissipation processes	conversion of friction to heat
distillation	important separation method in chemical technology, in which liquid or liquefied substances are separated out of other substances by evaporation and recondensation
elongation	the change in the length of a body pulled in one direction by the application of a force
exothermal reaction	a chemical reaction in which heat is released
filament	a continuous thread of a specified diameter, e.g. silk (natural fiber) or rayon (synthetic fiber); the opposite of a fiber with a finite length, such as wool or cotton
flow temperature (FT)	temperature above which a thermoplastic can be reshaped with little force
fluidized bed process	powdered or fine-grained material (e.g., quartz sand) is swirled upward by rising gases at a certain characteristic flow rate in such a way that the system resembles a liquid with respect to many of its properties; when used in the pyrolysis of plastics, this method allows heat to be transferred rapidly and the process to be performed in closed reactors
FRP	fiber-glass-reinforced plastic is a composite material consisting of glass fibers in a polymer matrix
functional groups	groups of atoms that impart a certain reactivity to molecules; functional groups also make it possible to distinguish material classes with corresponding chemical properties (e.g., hydroxyl groups of the alcohols, carboxyl groups of the organic acids, amino groups of the amines)
gel coat	resin layer (usually colored) that protects the underlying resin/glass fiber laminate from external influences such as impact, UV light, chemicals, etc.; the side of the molded part that remains most visible or external after demolding, hence the gel coat is the first layer applied to the mold
granulate	plastic starting material for primary processing; particles are usually in the form of lentil-shaped cylinders
hardener	second chemical component required for activating the cross-linking reaction of prepolymers, in order to manufacture thermosets or elastomers

holding pressure	conveys additional melt into the molded part as it solidifies in the injection molding process, thus minimizing shrinkage (loss of volume) that would otherwise occur as the injection-molded part cools; it also compresses the internal structure of the part
hydraulic	operated by the pressure of a liquid
injection molding cycle	represents the total of the durations of all operations performed within the injection molding machine and necessary to manufacture a part
injection pressure	pressure applied by the screw against the molding compound to push it into the mold in the injection process
investment	long-term provision of capital to replace used means of production (e.g. machines) and to procure new ones
ion exchangers	inorganic and organic substances that can exchange their own ions for others without undergoing a change in stability; used to soften (desalinate) water, for example
isotropy	condition in which properties are completely independent of direction (isotropic), i.e., the same in all directions
laminate	the cured thermoset matrix or the cooled fiber-reinforced composite (with a thermoplastic matrix)
layered structure	structure and arrangement of the individual layers (laminate) of a fiber-reinforced composite
locking force	force required to close the mold while being filled or during the hardening phase (in the case of thermosets)
macromolecular materials (polymeric materials)	materials consisting of threadlike or three-dimensional giant molecules with at least 1000 atoms; aside from synthetic materials, some natural materials, such as cellulose, proteins, and rubber, are included
matrix material	material that binds the fibers
melt	molten molding compound
modulus of elasticity (E-modulus)	constant ratio of stress to deformation within the elastic range of a material, determined by tensile test, compression test, and bending test; because of the viscoelastic characteristics of plastics, dependence on time is an important consideration
molded part	plastics part produced by primary processing, which is often possible to use without subjecting it to a finishing operation
molding compound	an unformed or preformed material that can be processed by noncutting methods of shaping (primary processing) within a certain temperature range to form a molded material, which may be a molded part or semifinished product

molecule smallest unit of a chemical compound

monomer basic building block from which macromolecules are created; for example, ethylene is the monomer of polyethylene ("monomer" is derived from the Greek word for "single part")

multifunctional multiple functions combined in a single component; for example, a cable on a ceiling light may bear the weight of the lamp itself while also conducting the electrical energy to the lamp

nonwoven nonwoven, solid, flat sheet consisting of bundled glass filaments or glass staple fibers (surfacing mat)

orthotropy also known as orthogonal or rhombic anisotropy, it indicates that properties depend on direction; the properties are symmetrical in relation to a system of three perpendicular (orthogonal) planes

PC polycarbonate (an amorphous thermoplastic)

PE polyethylene (a semicrystalline thermoplastic)

PEEK polyether etherketone (a semicrystalline thermoplastic)

PES polyether sulfone (an amorphous thermoplastic)

petrochemistry collective term for industrial, chemical, or physiochemical conversions of petroleum as the basic raw material

plasticizers substances that bring about softening of plastics; in a physical sense, plasticizing means decreasing the glass transition temperature (T_g) of a polymeric material, generally to a point below room temperature so that plastics that are initially hard, rigid, and brittle become soft, flexible, and impact resistant

plasticate (plastify) describes the conversion of a plastic material to a thermoplastic condition by the introduction of heat

PMMA polymethyl methacrylate (an amorphous thermoplastic)

polarity (of plastics) formation of electrical charge distributions within the macromolecules produces various polarities

polyaddition (step addition) chemical reaction in which the reactive groups or ends of monomers react with one another to form polymers, involving the migration of hydrogen atoms (transposition)

polycondensation (step polymerization) similar to polyaddition, except that water or another low-molecular substance is split off in the reaction; involves no migration of atoms or atomic groups, however

polymers long molecular chains (chain molecules) formed from monomers occurring as repetitive "building blocks" or monomer units in these chains ("polymer" is derived from the Greek word for "many parts")

POM	polyoxymethylene (a semicrystalline thermoplastic), also known as polyacetal
prepregs	molding compounds consisting of flat or linear glass reinforcements preimpregnated with thermosetting resin compounds; molding compounds prepared in this way are mostly glass fiber mats or glass filament cloths processed to form molded parts or semifinished products by hot-press molding
PS	polystyrene
PVC	polyvinyl chloride (an amorphous thermoplastic)
pyrolysis	thermal decomposition of chemical compounds
quasi-isotropic	a condition in which properties are nearly identical in all directions; in fiber-reinforced composites; this condition can be attained by providing at least three directions with reinforcement in similar layer thicknesses
raw material	naturally occurring starting material used to manufacture a product, such as petroleum, coal, ores, wood, hides, and cotton, as well as water and air; the intermediate product (semifinished product) is formed from the raw material during the production process, and the finished (manufactured) products are produced from the intermediate product
recycling	reutilization of raw or used materials; for example, plastic sprues from injection-molded parts are recycled by processing them to form a granulate, which is then returned to the injection molding process
refining	purification and improvement of natural materials and technical products (sugar, petroleum, etc.), performed in a refinery
resin	amorphous material with a consistency that can vary from soft to rigid; thermosetting resins form the basis for thermoset plastics
resin injection process	a closed mold is used to manufacture molded parts from resin, and reinforcing materials are inserted into the molded parts
RIM	reaction injection molding, which refers to an integrated mixing and injection process for plastics from two or more highly reactive components
roving	refers to a certain number of approximately parallel glass strands combined to form a larger strand (or glass roving); an individual glass strand consists of a certain number of individual glass filaments that have been combined without twisting to form a thread of uniform size, and these filaments are arranged in a mostly parallel relationship
sandwich	flat multilayer composite design consisting of two high-strength outer layers and a light, thick inner layer; this design provides a high surface moment of inertia and great stiffness

sealing point	time at which the molding compound inside the sprue channel has become solid enough to prevent any further flow
secondary valence forces	intermolecular forces with a very limited range of a few nanometers, e.g. hydrogen bonds
self-extinguishing	ability of a burning plastic to extinguish itself without external influence
semifinished product	intermediate product made of plastic (e.g., pipes and sheets), which will be further processed (reshaped) to form a finished product
shear strength (interlaminar)	property defined as the force sufficient to cause a rupture within the area subjected to shear divided by the area subjected to shear
softening temperatutre (ST)	temperature at which amorphous parts of a thermoplast melt
sonotrode	welding tool used in ultrasonic welding to transmit vibrations to the plastics part being welded
sprue bush	part of an injection mold that lies against the nozzle of the injection unit and through which the molding compound flows into the mold
stabilizers	chemical additives that make a plastic more resistant to certain influences (e.g., UV radiation, heat, oxidation, weathering)
synthesis	formation of chemical compounds from the basic elements or basic chemicals with a simpler structure ("synthesis" is derived from the Greek word for "put together")
thermoplastics (semicrystalline)	thermoplastics displaying crystalline and amorphous regions; plastics that can be melted (softened) by application of heat
thermoset	polymer in which the chain molecules are cross-linked in three dimensions through covalent bonds
ultimate elongation	the elongation that the body exhibits before breathing at the maximum amount of force; specified as a percentage of the starting length
unidirectional	aligned in one direction
UP	unsaturated polyester resin
viscoelastic	condition of a body that is both elastic (Hookean) and viscous (Newtonian)
vulcanization	chemical cross-linking process that alters the properties of natural rubber to make it elastic and resistant to distortion

Appendix II

Selected Literature

Baird, D. G. and Collias, D. I. Polymer Processing Principles and Design. Wiley, 1998

Beall, G. L. Rotational Molding. Hanser Publishers, 1998

Belofsky, H. Plastics: Product Design and Process Engineering. Hanser Publishers, 1995

Berins, M. L. (Ed.) Plastics Engineering Handbook of the Society of the Plastics Industry, 5th Ed. van Nostrand Reinhold, 1991

Birley, A. W. et al. Physics of Plastics. Hanser Publishers, 1992

Bisio, A. and Xanthos, M. How to Manage Plastics Waste: Technology and Market Opportunities. Hanser Publishers, 1994

Braun, D. (Ed.) Simple Methods for Identification of Plastics, 4th Ed. Hanser Publishers, 1999

Brown, R. P. (Ed.) Handbook of Polymer Testing. Physical Methods. Marcel Dekker, 1999

Chanda, M. and Roy, S. K. Plastics Technology Handbook, 3rd Ed. Marcel Dekker, 1998

Charrier, J.-M. Polymeric Materials and Processing. Hanser Publishers, 1990

Ehrig, R. J. (Ed.) Plastics Recycling: Products and Processes. Hanser Publishers, 1992

Gächter, R. and Müller, H. (Eds.) Plastics Additives Handbook, 4th Ed. Hanser Publishers, 1993

Gent, A. N. (Ed.) Engineering with Rubber, 2nd Ed. Hanser Publishers, 2000

Glenz, W. (Ed.) A Glossary of Plastics Terminology in 5 Languages, 4th Ed. Hanser Publishers, 1998

Gruenwald, G. Plastics: How Structure Determines Properties. Hanser Publishers, 1992

Hensen, F. (Ed.) Plastics Extrusion Technology, 2nd Ed. Hanser Publishers, 1997

Hofmann, W. Rubber Technology Handbook. Hanser Publishers, 1989

Holden, G. Understanding Thermoplastic Elastomers. Hanser Publishers, 2000

Holden, G., Legge, N. R. and Quirk, R. P. Thermoplastic Elastomers, 2nd Ed. Hanser Publishers, 1996

Johannaber, F. Injection Molding Machines, 3rd Ed. Hanser Publishers, 1994

Jones, R. F. Guide to Short Fiber Reinforced Plastics. Hanser Publishers, 1998

Lee, N. C. Blow Molding Design Guide. Hanser Publishers, 1998

Lee, N. C. Understanding Blow Molding. Hanser Publishers, 2000

Macosko, C. W. RIM Fundamentals of Reaction Injection Molding. Hanser Publishers, 1989

Malloy, R. A. Plastic Part Design for Injection Molding. Hanser Publishers, 1994

Menges, G. and Mohren, P. How to Make Injection Molds, 2nd Ed. Hanser Publishers, 1993

Michaeli, W. Extrusion Dies for Plastics and Rubber, 2nd Ed. Hanser Publishers, 1992

Michaeli, W. Plastics Processing: An Introduction. Hanser Publishers, 1995

Michaeli, W. et al. Training in Injection Molding. Hanser Publishers, 1995

Odian, G. Principles of Polymerization, 3rd Ed. Wiley-Interscience, 1991

Osswald, T. Polymer Processing Fundamentals. Hanser Publishers, 1998

Osswald, T. and Menges, G. Material Science of Polymers. Hanser Publishers, 1995

Poetsch, G. and Michaeli, W. Injection Molding – An Introduction. Hanser Publishers, 1995

Progelhof, R. C. and Throne, J. L. Polymer Engineering Principles. Hanser Publishers, 1993

Rao, N. S. Design Data for Plastics Engineers. Hanser Publishers, 1998

Rauwendaal, C. Extrusion, 3rd Ed. Hanser Publishers, 1994

Rauwendaal, C. Understanding Extrusion. Hanser Publishers, 1998

Rauwendaal, C. Polymer Mixing. Hanser Publishers, 1998

Rees, H. Understanding Injection Molding Technology. Hanser Publishers, 1994

Rees, H. Understanding Product Design for Injection Molding. Hanser Publishers, 1996

Rees, H. Mold Engineering. Hanser Publishers, 1995

Rosato, D. V. and Rosato, D. V. (Eds.) Blow Molding Handbook. Hanser Publishers, 1989

Rosato, D. V. and Rosato, D. V. (Eds.) Injection Molding Handbook, 2nd Ed. Kluwer Academic Publishers, 1994

Rosato, D. V. Rosato's Plastics Encyclopedia and Dictionary. Hanser Publishers, 1992

Rosen, S. L. Fundamental Principles of Polymeric Materials, 2nd Ed., Wiley-Interscience, 1993

Rotheiser, J. I. Joining of Plastics. Hanser Publishers, 1999

Rubin, I. I. (Ed.) Handbook of Plastic Materials and Technology. Wiley-Interscience, 1990

Saechtling, Hj. (Ed.) International Plastics Handbook, 3rd Ed. Hanser Publishers, 1995

Selke, S. E. M. Understanding Plastics Packaging Technology. Hanser Publishers, 1997

Selke, S. E. M. Plastics Packaging. Hanser Publishers, 2000.

Seymour, R. B. and Carraher, C. E. Giant Molecules. Wiley-Interscience, 1990

Stoeckhert, K. and Mennig, G. Mold-Making Handbook, 2nd. Ed. Hanser Publishers, 1999

Throne, J. L. Thermoplastic Foams. Sherwood Publishers, 1996

Throne, J. L. Understanding Thermoforming. Hanser Publishers, 1999

Todd, D. B. Plastics Compounding. Hanser Publishers, 1998

Tres, P. A. Designing Plastic Parts for Assembly, 4th Ed. Hanser Publishers, 2000

Uhlig, K. Discovering Polyurethanes. Hanser Publishers, 1999

Ulrich, H. Introduction to Industrial Polymers, 2nd Ed. Hanser Publishers, 1992

White, J. L. Twin Screw Extrusion: Principles and Practice. Hanser Publishers, 1990

White, J. L. Rubber Processing: Technology, Materials, and Principles. Hanser Publishers, 1995

Wildi, R. H. and Maier, Chr. Understanding Compounding. Hanser Publishers, 1998

Wright, R. E. Molded Thermosets. Hanser Publishers, 1991

Wright, R. E. Injection/Transfer Molding of Thermosets. Hanser Publishers, 1995

Xanthos, M. (Ed.) Reactive Extrusion. Hanser Publishers, 1992

Appendix III

Example of a Training Program for Plastics Processing Operators in the US

Certification – the Next Step to a Well Trained Workforce

The plastics industry has seen a tremendous growth over the last 75 years to become the fourth-largest manufacturing segment in the United States. As a result of this rapid growth, employers have encountered problems hiring qualified personnel. In addition to this shortage of qualified personnel there is also a need for an agreed upon set of knowledge and skill standards within the plastics industry.

Members of the plastics industry recognized the importance of defining an industry standard. Establishing a national standard makes it easier for employers and educators to identify the job-related knowledge, skills and abilities that are critical to attracting, training and retaining productive workers. In addition it can be used as a benchmark against which employers and supervisors can measure an operators level of knowledge as compared to an agreed-upon industry standard and guide curriculum development. The Society of the Plastics Industry Inc. (SPI) was given the mandate, by its members, to research and create a standard that could be recognized by the industry at large.

SP recruited the assistance of the plastics industry to develop the standard. From an extensive job analysis of nearly 5,000 machine operators and supervisors in 421 thermoplastics manufacturing facilities in the United States SPI developed what is now the defined industry standard.

The National Certification in Plastics (NCP) standard focuses on seven key areas: safety; quality assurance; basic process control; general knowledge, preventive and corrective action on primary and secondary equipment; handling, storage, packaging and delivery of plastics materials; and tools and equipment.

The standard was used to create the *National Certification in Plastics Certified Operator* exam for machine operators. The NCP exam is the first-ever exam to test machine operators on the industry-wide standard in all four major plastics processes—injection molding, blow molding, extrusion and thermoforming. The exam uses the seven content areas of the standard to test the skills and knowledge identified by the industry as being important.

The major knowledge content areas, subcontent areas and relative weights of importance covered on the NCP exam are:

I. Basic Process Control (16%)
II. Preventive and Corrective Action on Primary/Secondary Equipment (12%)
III. Handling Storage, Packaging, Preservation, Delivery of Materials (11%)
IV. Quality Assurance (18%)
V. Safety (21%)
VI. Tools and Equipment (8%)
VII. General Knowledge (14%)

Examination results are delivered to operators upon completion of the exam. Operators achieving certification will be given an unofficial report congratulating them on their success. Unsuccessful operators are given a diagnostic report that details their performances vs the standard for each content area. These reports can be used to identify areas of improvement in individuals and gaps that may be in the training programs if several people from a company are unsuccessful.

Definition of an NCP Certified Operator

The plastics machine operator who meets the standard of *productive performance* will demonstrate the knowledge necessary to operate the machine, auxiliary equipment, and related tools safely and to know what to do in emergency situations. He/she must understand the quality parameters and data collection required for a given product, must be able to recognize when the machine is malfunctioning or the process is out of control, and must be able to initiate the corrective action required. He/she must have a *basic* knowledge of processing including setup, start up/shut down, and standard documentation. He/she must also have *basic* math and reading comprehension skills, as well as a *basic* knowledge of the handling, storage, packaging, and delivery of plastics materials in the plant. He/she will be familiar with team-oriented techniques.

Proper training for operators is widely acknowledged as important, but few companies do it. The challenge is figuring out what training is necessary and how to provide it. The NCP Body of Knowledge, based on the industry standard, is an outline of the content and subcontent areas covered on the exam. Various educators and companies in the industry have taken it and used it to create curriculum.

Basic Process Control

Knowledge of...

A. Operations

1. Machine Operations
   - ☐ process flow from raw material to finished product

B. Manufacturing Knowledge

1. Team Building and Work Group Techniques
   - ☐ information sharing
   - ☐ meeting participation
   - ☐ team participation
   - ☐ achieving consensus and compromise
   - ☐ goal setting
2. Time Management Techniques
   - ☐ organization
   - ☐ planning
3. How to Initiate Changes for Quality Improvement
   - ☐ implementing new procedures for performing tasks
   - ☐ process improvements
   - ☐ documenting and communicating improvement ideas
4. General Manufacturing Practices
   - ☐ standards
   - ☐ policies/procedures
   - ☐ work instructions
5. How Defects Affect Final Product
   - ☐ customer dissatisfaction

Example of How to Develop a Training Plan

Establish the Scope of Your Operator Training Program

- Establish entry level requirements
  - First phase: foreman, lead operators, orientation operators, or anyone who currently does your new operator training
  - Second phase: senior employees, operators who will be advancing soon
- Decide how many will complete the initial phase of training
- Decide how training will be delivered
- Classroom
- OJT
- Interactive self-study
  - Establish hours of required employee involvement
  - Establish cycle time frame from start to completion of training
  - Establish company goals for operator training

Establish "Body of Knowledge" Presented in Training Program

- Collect material needed to create training curriculum
  - NCP Body of Knowledge
  - Training manuals
  - Machine manuals
  - Videos
  - Safety literature
  - Interactive computer training

  - NCP sample exam on disk
  - Speak with operators currently doing the job who are knowledgeable
- Edit material collected by asking yourself:
  - Will our Operators use any of these skills, knowledge or attitudes in the next 48 hours to do their job? This material will become your first phase of training.
  - Will Operators occasionally be involved with activities where additional skills, knowledge or attitudes would make them more valued employees? This material becomes the second phase of training for operator advancement.
  - Will this information give them a greater understanding of what their co-workers in the shop are doing even though it doesn't involve them directly? This material will also become part of the second phase of training for operator advancement.
  - Nice to know, but operator job as defined in your shop doesn't get involved in these issues. Defines the additional issues that may be useful for certification.
- Establish milestones in curriculum
  - Probation—how much of phase 1 should operators know
  - Review—have they finished phase 1 topics
  - Certification—preparing for NCP exam

Resolve Logistical Issues in Your Training Program

- Establish where training will occur
  - Keep in mind how adults learn when deciding amongst a classroom setting inside or outside, on the shop floor, at a computer, combination of those things etc.
- Determine who will do the training or document progress
  - Training Coordinator
  - Human Resources
- Establish how the employees will be able to attend/participate on a regular basis
- Order training materials for Trainer/Facilitator
- Order training materials for operators
- Establish a means of communicating training schedule to employees

Resolve Human Resource Issues

- Establish the operator career path
  - By defining a career path your company says to it's operators they are skilled participants in a fast growing, technically demanding, sector of manufacturing
- Establish a linkage of training program to career path
- Establish a visible tracking method for training progress
- Establish rewards progress
  - Monitory
  - Recognition
  - Promotion
- Establish means to test the effectiveness of training program
  - NCP exam

- Establish companies' policy for cost of certification

Post Training Activities

- Establish if your program is on-going or periodic
- Establish how your training can be delivered one on one or establish how many new operators will make a new class
- Establish what needs to be changed in your program to improve or update your topics or materials

To find additional resources for training visit the NCP website at www.certifyme.org for a list of trainers and educators who have incorporated the NCP Body of Knowledge in their program.

NCP is committed to addressing the needs of the industry regarding workforce development. The current NCP standard serves four manufacturing processes and covers the machine operator job level. NCP is working toward addressing the need for certification at the technician job level and intends to have a standard available in the future.

For information on the certification program for machine operators visit the NCP website at www.certifyme.org or send an email to ncp@socplas.org. Information is also available by contacting +1 (202) 974-5356.

Answers to Review Questions

Lesson 1

1 thermosets
2 semicrystalline
3 melt
4 insoluble
5 lightly
6 swellable
7 thermoplastics
8 lighter
9 lower
10 different
11 good
12 can

Lesson 2

1 natural gas
2 cracking
3 propylene
4 chain
5 polymer
6 carbon (C)
7 poly
8 tangled
9 carbon (C)

Lesson 3

1 double
2 "coupling"
3 copolymers
4 polypropylene (PP)
5 splitting off
6 water
7 two or more
8 polycarbonate (PC)
9 "splitting off"
10 functional groups
11 epoxies
12 "exchange"

Lesson 4

1 atomic bonds
2 intermolecular bonds
3 stronger

Lesson 5

1 semicrystalline
2 glassy
3 densely
4 swell
5 transparent
6 repeatedly

Lesson 6

1 elongation at break
2 strength
3 ductility
4 +50°C (122°F)
5 below
6 −15°C (60°F)
7 amorphous
8 rigidity
9 cross-linked
10 +130°C (265°F)

Lesson 7

1 strength
2 1000
3 is
4 creep
5 heated
6 orientation
7 time and temperature
8 time-dependent creep diagrams
9 5
10 50

Lesson 8

1 lighter
2 0.9–2.3
3 2000
4 metal powder
5 roughly equal to
6 transparency

Lesson 9

1 primary processing
2 reshaping
3 cementing
4 milling
5 thermoforming
6 welding

Lesson 10

1 processing
2 mixers
3 weight
4 plasticizing
5 more freely than
6 cutting mills

Lesson 11

1. continuously
2. the extruder
3. three-zone screw
4. high
5. shape
6. multiple layers
7. extrusion blow molding

Lesson 12

1. primary processing
2. mass-produced parts
3. cycle time
4. finished parts
5. mold
6. shrinks
7. cooling period
8. screw

Lesson 13

1. matrix
2. 490,000 N/mm^2 (71,050,000 psi)
3. 
4. (b) impregnation; (c) shaping; (d) curing
5. hand lay-up
6. (a) thermosetting; (b) thermoplastic

Lesson 14

1. gas bubbles
2. lighter than
3. uniformly
4. higher
5. false
6. foaming methods
7. high
8. injection molding

Lesson 15

1. heated
2. thermoplastics
3. infrared radiation
4. only one side
5. prestretched
6. shorter than

Lesson 16

1. molten
2. cannot
3. miscibility, viscosity and melt temperatures
4. thermal conduction
5. hot gas
6. internal and external
7. complicated

Lesson 17

1 less
2 thermal expansion
3 smaller than
4 can
5 highest
6 liquid

Lesson 18

1 can
2 cohesion and adhesion
3 clean
4 peeling
5 thermosets
6 quickly

Lesson 19

1 short
2 long
3 long
4 45
5 a lot of
6 hardly degrade
7 more

Lesson 20

1 possible
2 feedstock recycling
3 energy and raw materials
4 can be recycled
5 grained
6 pure
7 unmixed
8 high
9 waxes